Designing the Sustainable Site

HEATHER VENHAUS

Designing the Sustainable Site

INTEGRATED DESIGN STRATEGIES FOR SMALL-SCALE SITES AND RESIDENTIAL LANDSCAPES

John Wiley & Sons, Inc.

This book is printed on acid-free paper. ♾

Published by John Wiley & Sons, Inc., Hoboken, New Jersey.

Published simultaneously in Canada.

Library of Congress Cataloging-in-Publication Data

Venhaus, Heather.

Designing the sustainable site : integrated design strategies for small-scale sites and residential landscapes / Heather Venhaus.

p. cm.

Includes bibliographical references and index.

ISBN 978-0-470-90009-3 (pbk.); ISBN 978-1-118-18026-6 (ebk); ISBN 978-1-118-18027-3 (ebk); ISBN 978-1-118-18341-0 (ebk); ISBN 978-1-118-18342-7 (ebk); ISBN 978-1-118-18343-4 (ebk)

1. Homesites—Planning—Case studies. 2. Sustainable design—Case studies. 3. Ecological landscape design—Case studies. I. Title. II. Title: Integrated design strategies for small-scale sites and residential landscapes.

NA9051.V46 2012

720.28—dc23

2011033592

Printed in the United States of America

SKY10091328_111924

For Emmett and all the wonderful little friends you have brought into our lives.

Watching you explore this beautiful world is one of our greatest pleasures. We can't wait to see all the amazing things you will accomplish.

Contents

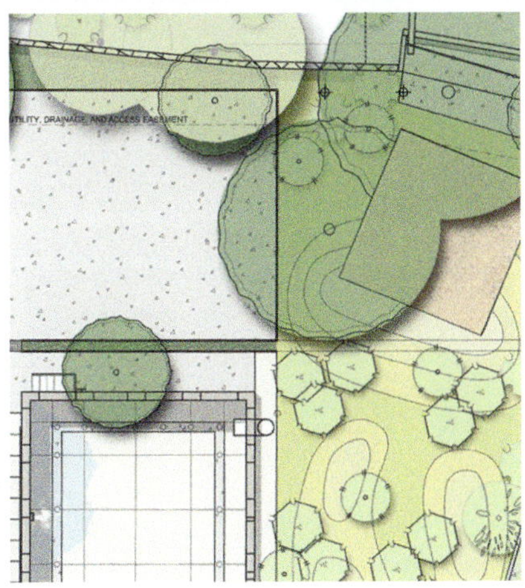

taylor28

Foreword

HUMANITY'S SUCCESSFUL QUEST for sustainability relies upon our connection with nature and the understanding that healthy, functioning ecosystems are directly related to our survival. Throughout the 30 years of my professional career, I have noticed a dramatic increase in environmental problems, but I have also noticed an increase in the willingness of people to work for a better world.

In this modern era, we are continually confronted with serious problems that are pertinent to our very existence—climate change, water shortages, and loss of biodiversity—and I believe that there is no time to delay corrective action or to accept easy options. In order to build a sustainable future, we must look within ourselves. We must understand the impact of our actions, face tough challenges, and seize the opportunities that are in front of us. Most importantly, we must act upon what surrounds us—our cities and the landscapes that we encounter on a daily basis.

It is often the small examples that make change possible and that have a great impact on new movements in society, politics, and architecture. They help us develop new standards, give us hope, and provide us with the courage to redirect our actions. Knowing this, Heather Venhaus has chosen to focus on such small-scale sites and residential landscapes as schoolyards, parks, residential developments, backyards, and streetscapes that people encounter regularly and that therefore have a significant impact on their well-being and quality of life.

Designing the Sustainable Site is more than just another book for the bookshelf. It is a call to action and a guide to observing and working with the built and natural environments in order to achieve sustainability right in our own cities and neighborhoods. In the pages that follow, readers will find the tools necessary to work holistically, to restore ecosystem function, and to develop high-performance sites that celebrate beauty and incite emotional energy. It is my hope that this book will have a large audience, as it is a strong contribution toward livable cities and an environment that we all need today and for the future.

—Herbert Dreiseitl

Preface

OVER 7 BILLION PEOPLE now inhabit the earth, placing unprecedented pressure on the planet's soils, waters, forests, and other natural capital. The majority of the global population lives in urban areas, where their interactions with nature, and the benefits that these interactions provide, commonly occur in small-scale sites and residential settings. Most often, these landscapes are treated as inconsequential, and their full potential to mend humanity's environmental offenses and improve our quality of life is commonly overlooked. This book was written to address this issue and to assist projects in gaining the full environmental, economic, and social benefits that can be achieved when sites protect and restore ecosystem services. It seeks to elevate the discussion of sustainability beyond "doing less bad"—attempting to merely slow down environmental degradation—to create regenerative sites that restore ecosystem function and rebuild the earth's natural capital.

This book explores major environmental and human health issues, such as air and water pollution, habitat loss, water shortages, and flooding, which often plague urban environments, as well as the potential for site development and maintenance to either contribute to these problems or to be part of the solution. Sustainable strategies that address each challenge include detailed descriptions, design considerations, and illustrations to help project teams determine the best options for their site. Throughout, the book emphasizes the interconnectivity of all project components and helps designers integrate living and built systems into mutually beneficial and cohesive design solutions. Integrated design is stressed as a model for improving site performance and saving time and money over the life of the project.

All sites—whether densely urban, suburban, or rural—can support the natural systems and processes that sustain and fulfill our lives. Throughout the book, numerous case studies from public and private projects in the United States and abroad are provided to illustrate a diversity of sustainable design strategies. These projects demonstrate that sustainability happens, not in spite of but in response to challenges. As with all projects, the design teams for the case studies faced outside influences, budget limitations, and other restrictions, but through focused effort, creativity, and collaboration, they were able to create sustainable solutions. In many cases, these projects are more cost effective and provide a broader suite of ecosystem services than similar conventional landscape developments. It is my hope that the case studies will motivate the design and landscape industries to continue raising the bar and striving toward true sustainability.

Acknowledgments

A NUMBER OF PEOPLE'S encouragement, generosity, and guidance have made this book possible. In particular, I would like to thank the committee members and staff of the Sustainable Sites Initiative. The five years I spent working with you defining site sustainability and developing the Guidelines and Performance Benchmarks was an awesome and rare learning opportunity. Your dedication and vast knowledge is truly inspiring. The memories of our many meetings, literature reviews, insightful discussions, and lively debates were my constant companion while writing this book.

I am grateful to the photographers, designers, and firms who so graciously lent their work to help illustrate the beauty and important message of sustainable site design. My gratitude goes out to Dr. Chris Martin, Emily Manderson, Steven Rodie, Dr. Mark Simmons, Jacob Blue, Meg Calkins, and Dr. Nina Bassuk, who provided review and feedback of draft manuscripts, and to my editor, Margaret Cummins, and Senior Production Editor Donna Conte, for their patience and friendly approach throughout the process. I would like to acknowledge the Austin Public Library and the University of Texas libraries' staff and interlibrary loan offices—an invaluable resource that is often underappreciated—who cheerfully helped me to quickly access a wide variety of books and journals from their libraries and others across the United States.

I am especially thankful to my family and friends for their ongoing interest, support, and humor throughout the development of this book, and to my son Emmett for taking long naps (most of the time) and for providing me with lots of entertainment and healthy distractions along the way. Finally, I am grateful to my husband, friend and personal ecologist Matt McCaw, whose steady belief in my abilities continually refuels my energy and my confidence. His patience, scientific insight, and meticulous editing helped me to stay focused, improve content, and see the book to the end.

—Heather Venhaus

I AM IMMENSELY PROUD to have worked on this book for many reasons, chief among them because it chases down sustainability and seeks to crystallize it for the reader as an absolute. Sustainability is not a gradient. Our endeavors are either sustainable or they aren't. From here, this book shows the reader that sustainability is not only necessary but also entirely possible, and indeed beginning to materialize. I must thank Heather for giving me the opportunity to contribute to this project and sign on to its message. Heather had to seize the moment with this book, so the timeline was much compressed and the writing process intense. Nevertheless, it was a pleasure—and a privilege. Looking back, one never misses the sleep he's lost....

I am also immensely grateful to the folks who helped make my little piece of this project happen. Jean Krejca and Kellie Cowan, Linda Haugen, Steve Dewey, Mark Simmons, Erica Nix, Casey Boyter, Mrs. W. D. Bransford, and Joe Marcus so graciously made brilliant photos available. Many thanks also to Damon Waitt, Travis Gallo, and Alison Arnold for their expert guidance and reviews. Special thanks are due to Amy Bellaire, whose thorough and honest critique and counsel significantly improved the quality of the manuscript and helped to clarify some of my own thoughts on biodiversity.

Lastly, the Austin Public Library's on-line research tools were absolutely critical to this endeavor. The APL and their dedicated staff are such a valuable resource to researchers, writers, and practitioners.

—Matt McCaw

■ FIGURE 1.1
Pudong, the financial district of Shanghai, was primarily farmland and countryside prior to 1993. Today the 467-square-mile Chinese district has a population over 5 million and a density of 10,794 people per square mile (4,168 people per km^2).

CHAPTER 1

Building a Sustainable Future

CHANGE OCCURRED RAPIDLY in the twentieth century—more so than at any other time period in the history of humanity. Arguably, the most significant change has been the number of people living on earth and depending on its resources for survival. Within a hundred-year time span, the global population grew from 1.6 to 6 billion, and for the first time in history over 50 percent of the population—80 percent in the United States and Europe—is concentrated in urban areas. Cities are hastily expanding to accommodate the rapid influx. In the United States alone, 1.5 million acres (0.6 million hectares) of farmland, forest, or other rural land is being converted to urban development each year (American Farmland Trust 2009). In the coming decades, the rapid population increase is expected to continue, with projections of 7 billion in 2011, 8 billion in 2024, and 9 billion by 2045.

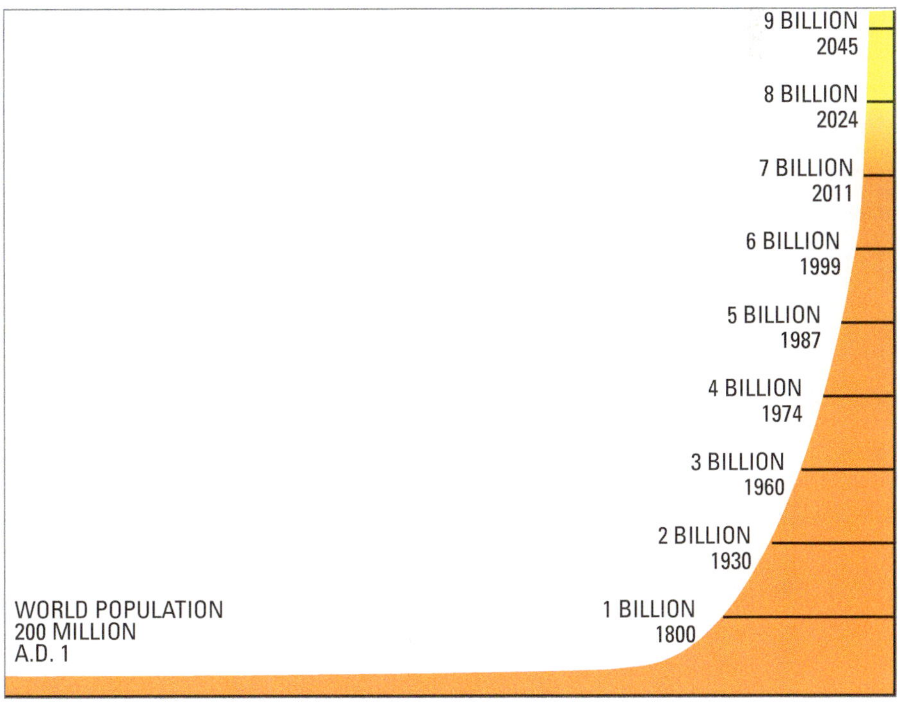

FIGURE 1.2
Global population growth.

As human populations increase, so do the demands on the earth's resources. Unprecedented pressure is being placed on the planet's soils, waters, forests, and other natural capital (Brundtland 1987). It is projected that at current rates, humanity will soon need the capacity of two earths to absorb CO_2 waste and keep up with natural resource consumption (World Wildlife Fund 2010).

To maintain their physical and mental health, every individual needs and *deserves* clean air, clean water, healthy productive soils, opportunities for physical activity and mental respite, and other benefits or "ecosystem services" provided by the natural environment. Historically, we have not required urban sites to function as sustainable and productive ecosystems but instead have relied on wildlands or rural areas to provide the services that sustain human life. Sadly, two-thirds of ecosystem services are now in decline worldwide (UN Foundation 2005).

Urban sites and other developed landscapes can help reverse this trend. A sustainable future for the growing population is not out of reach, but achieving it will require dramatically changing the ways in which sites are developed and maintained. To adequately provide for the next generation, the protection and restoration of ecosystem services must become standard practice for all sites—both urban and rural.

■ ECOSYSTEM SERVICES: A KEY ATTRIBUTE OF A SUSTAINABLE SITE

Ecosystems provide a multitude of resources and processes that sustain and fulfill human life. These benefits, collectively known as ecosystem services, are essential to our well-being and are a key attribute of a sustainable site. Examples of ecosystem services include:

- Regulate temperature and precipitation.
- Sequester greenhouse gases.
- Cleanse the air and water.
- Provide habitat.
- Maintain soil health and fertility.
- Retain and store fresh water.
- Control erosion.
- Provide recreation.
- Recycle nutrients.
- Produce food and other raw materials such as timber, medicine, and fuel.
- Mitigate natural hazards such as flooding, wildfire, and drought.
- Provide inspiration, intellectual stimulation, and cultural enhancement.
- Enhance opportunities for mental respite.

Many of the goods and services provided by nature are often taken for granted, in large part because they are supplied for "free" and are not part of our traditional accounting systems. To underscore their importance and inform land-use decisions, scientists have begun estimating the wealth of ecosystem services and have found the monetary value to be an average of $33 trillion per year, or nearly twice the global gross national product (Costanza et al. 1997).

Issues that plague urban environments, such as flooding, urban heat islands, and water pollution, are often caused or exacerbated by the disturbance or removal of natural systems and the benefits they provide. Sustainable sites seek to improve the quality of life of site users and the surrounding communities by creating regenerative systems that protect and restore ecosystem services.

Regenerative Systems

The building industry has been an early adopter of the sustainability movement and has documented success in reducing energy, water use, greenhouse gas emissions, and solid waste. Although reducing environmental impacts is definitely a step in the right direction, it is not enough to provide a sustainable future for the burgeoning human population. In addition to doing less damage, we must also reverse the degradation of the earth's natural resources by creating regenerative and resilient systems that sustain and increase the provision of ecosystem services. Landscape practitioners can lead the green building movement to a higher level of sustainable design by helping project teams realize this goal and integrate living systems into all aspects of the site.

Previously developed sites that have limited ecological or cultural value present the greatest opportunity for the type of regenerative change we need. The redevelopment of environmentally degraded sites, such as greyfields or brownfields, provides a mechanism not only for protecting native ecosystems and agricultural lands (via diversion of development pressure) but also for restoring natural systems and the ecosystem services they provide. Encouraging development within existing communities and developed places also conserves the natural and financial resources required to construct and maintain infrastructure. This stands in contrast to the development of greenfield sites, which has a much greater potential of reducing or destroying healthy, functioning ecosystems and the goods and services they provide. Greenfield development that diminishes ecosystem services ultimately contributes to the global decline of natural capital and the overall benefits humanity receives from nature.

■ **FIGURE 1.3**
High Line Park, Twenty-sixth Street viewing spur. The elevated public park constructed on an abandoned railway in Manhattan repurposes existing structures and provides a ribbon of green space that restores a variety of ecosystem services in a dense urban environment.

What Is Site Sustainability?

Sustainable development is commonly defined as "development that meets the needs of the present without compromising the ability of future generations to meet their own needs" (Brundtland 1987). It recognizes the interdependency between the environment, human health, and the economy and considers all three when measuring success.

The three pillars of sustainability and their relationship to site development are outlined below:

- **PLANET:** Environmental or ecological sustainability stems from the realization that human life (and the life of other creatures as well) is dependent upon the natural environment and its provision of ecosystem services. It recognizes that there are limits to the bounty ecosystems can provide and to the harvest and degradation they can withstand. To ensure the longevity and viability of the earth's resources, sites must protect and restore ecosystem services and humans must act as stewards of the environment. Sustainable sites help society build an environmental ethic by providing everyday opportunities for people to connect with nature.
- **PROFIT:** Traditionally, the success of development has primarily been evaluated by economic measures. Placing such a strong focus on financial gains alone has led to significant environmental and human health costs. For any endeavor to work long term, it must certainly be profitable; however, other factors must also be considered. Sustainable sites base decisions not only on their economic merits but also on their environmental and social costs and benefits. Including the impacts on people and the planet in the project accounting brings to light the full cost of doing business and encourages more social and environmental responsibility.
- **PEOPLE:** Social equity and human health is an aspect of sustainability that is commonly overlooked and can be the most difficult to address. It extends the opportunity to aspire to a better quality of life to all individuals. Social equity addresses basic provisions such as clean air and water, the right to education, access to safe and healthy green space, and other factors that impact our quality of life. Sustainable sites play an important role in supporting human health and create opportunities for all site users to improve their physical, mental, and social well-being.

■ **TABLE 1.1**
Example Characteristics of Conventional and Sustainable Sites

	CONVENTIONAL SITE	SUSTAINABLE SITE
TEAM CULTURE OR PHILOSOPHY	Perceives nature and development as being in opposition. May incorporate sustainable practices into the design if it does not increase time or immediate costs.	Values nature and the ecosystem services it provides. Accepts the responsibility of sustainability and providing a meaningful quality of life to future generations. Strives to reverse the degradation of the earth's natural resources by creating regenerative and resilient systems.
MEASURES OF SUCCESS	Primarily evaluated by the economic success of the project.	Success is measured by not only the economic outcomes but also the environmental and human health impacts of the project.
DESIGN PROCESS	Site design is compartmentalized, and the landscape and buildings are viewed as separate entities. Landscape design often begins after the building design or construction is complete. Consultants work independently on their area of the project and communicate information as needed.	Building and landscape practitioners, engineers, construction and maintenance professionals, and other consultants are collectively involved in the design process and work together to optimize the performance of the site toward common goals.

	CONVENTIONAL SITE	SUSTAINABLE SITE
AESTHETICS	Somewhat homogenous, replicating standard templates similar to sites from other regions or parts of the world.	Design solutions grow from the place and are representative of the local soils, vegetation, materials, and culture.
ENERGY	Relies heavily on nonrenewable resources that harm the environment and human health.	Minimizes energy consumption and the use of fossil fuels. Whenever feasible, energy is derived from the sun and wind, biomass, or other renewable resources.
	The building and landscape do not work together to reduce energy consumption.	The landscape creates favorable microclimates that reduce the energy consumption of buildings and increase the comfort of site users.
SOILS	Construction and maintenance practices commonly damage soils.	The disturbance of healthy soils is minimized. Degraded soils are restored prior to replanting.
	Require regular applications of fertilizers to promote healthy plant growth.	Soil biota and organic matter from on-site vegetation promote healthy plant growth.
VEGETATION	Preserves large trees.	Maximizes the integration of all existing native and ecologically appropriate vegetation into the site design.
	Plant selection is primarily based on site conditions and aesthetic considerations.	Plant selection considers a broad range of factors, including growing conditions, beauty, resiliency, ecological function, native range and habitat, invasiveness, and maintenance requirements.
WATER	Quickly conveys stormwater runoff and other wastewater resources off-site.	Captures rain and wastewater for reuse on-site or on adjacent properties.
	Strongly relies upon potable water for irrigation.	Landscape primarily relies upon precipitation or wastewater resources such as air-conditioner condensate, greywater, or reclaimed water.
MATERIALS	Removes and disposes of much of the existing building and landscape materials.	Maximizes the reuse of existing structures, landscape, and building materials.
	The reuse of site structures or materials at the end of the project life is not considered in the design process.	Sites are designed to minimize the disposal of materials. Site structures and features can be adapted and reused in place or easily deconstructed and reclaimed or recycled.
MAINTENANCE	The individuals responsible for maintenance are not aware of the goals of the project or how maintenance practices impact the site's ecological and cultural function.	The individuals responsible for maintenance understand and support the goals of the project. Education and training is provided to ensure that maintenance optimizes the site's ecological and cultural performance.
	Maintenance occurs on a regular schedule and is not informed by the performance of the site. Land-care practices focus on keeping the site somewhat static and limiting change.	Postoccupancy evaluations and monitoring guide land-care practices. The site evolves and adapts in a way that continually improves its ecological function and the visitor's experience.
CONTINUED LEARNING	No postoccupancy evaluations or monitoring is conducted to improve future projects.	Monitoring is built into the design and information gathered is used to improve future projects and the success of the sustainable design industry.

The Importance of Education and Stewardship

Design alone cannot ensure a sustainable site; what is created on paper must be translated into a tangible project constructed and cared for in a way that perpetuates its success. Landscape practitioners often guide the design and construction process but are commonly separated from the long-term management of the site. Many project teams that have worked so diligently to minimize resource consumption, cleanse water, restore ecological processes, and address other aspects of sustainability discover after some time that their sustainable site does not function as intended or live up to its accolades. This is often due to a lack of performance monitoring and misguided or omitted operations and maintenance procedures. These important practices are frequently overlooked or cut from the project for one or all of following reasons:

- Budget restraints
- A belief that landscapes are natural systems and as such can care for themselves
- A lack of individuals who take ownership of the site and see themselves as stewards of the land
- A general ignorance or apathy toward the concept of sustainability and how the site must function in order to support it

Regardless of the reason, the fact stands true: constructed landscapes and many natural systems do require monitoring and strategic management and *stewardship* in order to continue to function properly and optimize their provision of ecosystem services. Accepting this, we must ask ourselves, how do we get people to embrace sustainability and care about the ecological health of our landscapes? How do we instill a sense of stewardship for our built and natural ecosystems? The answer is twofold: (1) illustrate both the short- and long-term economic and human health benefits, and (2) provide educational and meaningful experiences that connect people to nature. In addition to project teams working with the client, maintenance staff, or volunteers to help them understand why monitoring and stewardship are central to long-term success; project teams can also create landscapes that help humanity build an environmental ethic.

Aldo Leopold, in his writings on the subject, recognized the need for a land ethic—a moral principle or value—that "simply enlarges the boundaries of our community to include soils, waters, plants, and animals, or collectively: the land." Leopold notes, "An ethic to supplement and guide the economic relation of land presupposes the existence of some mental image of land as a biotic mechanism. We can be ethical only in relation to something we can see, feel, understand, love, or otherwise have faith in" (Leopold 1949).

■ **FIGURE 1.4**
Infiltration planters filled with trees, grasses, and perennial wildflowers manage stormwater and connect the surrounding community to the natural environment at the Taylor 28 streetscape.

In other words, nature must become relevant to our everyday lives. Proving opportunities for society to see and experience nature in this way should be the charge of landscape practitioners and one of the primary purposes of a sustainable site. In this respect, a successfully designed site functions as a living teaching tool.

There are many different ways to learn, and the best teachers make a topic relevant to their students. In a landscape setting this can be accomplished through both active education or outreach and experiential learning. In addition to commonly used conventional teaching methods such as interpretation, guided tours, or volunteer activities, landscapes can also teach by being a source of inspiration, evoking emotion and providing a physical connection to the environment. Constructed landscapes can reveal the ecological processes, rhythms, and cycles of nature and provide restorative settings that allow us to reflect upon our place in the world and to notice the environment around us (Meyer 2008). Hands-on interaction and exploration of diverse and healthy ecosystems can build a broader understanding of the natural world and provide the motivational basis for more formal learning (Wells and Lekies 2006). Positive and spontaneous interaction with nature in our homes, schools, and places of work can build a familiarity with and love for the natural environment that translates into a sense of stewardship. Landscapes that improve our understanding of nature and make it relevant to our lives can ultimately have a sphere of influence that extends well beyond the boundaries of the site. Though the number of people who visit the site may be relatively small in comparison to the global population, their environmental ethic can be very influential and a catalyst for change, impacting the government officials they elect, their vote on key issues, the purchase of products, and decisions on where to live and how to commute (Meyer 2008).

CASE STUDY

UNDERWOOD FAMILY SONORAN LANDSCAPE LABORATORY

PROJECT TYPE: Public institution

LOCATION: Tucson, Arizona

SIZE: 1 acre (0.4 hectare)

HIGHLIGHTED SUSTAINABLE PRACTICES:

- Redevelopment of a greyfield site
- Use of harvested wastewater
- Increased vegetative biomass
- Habitat for endangered species
- Landscape irrigation requirements balanced with the available wastewater supply
- Comfortable outdoor microclimate that encourages interaction with nature
- On-site monitoring and documentation of sustainable practices to evaluate performance over time

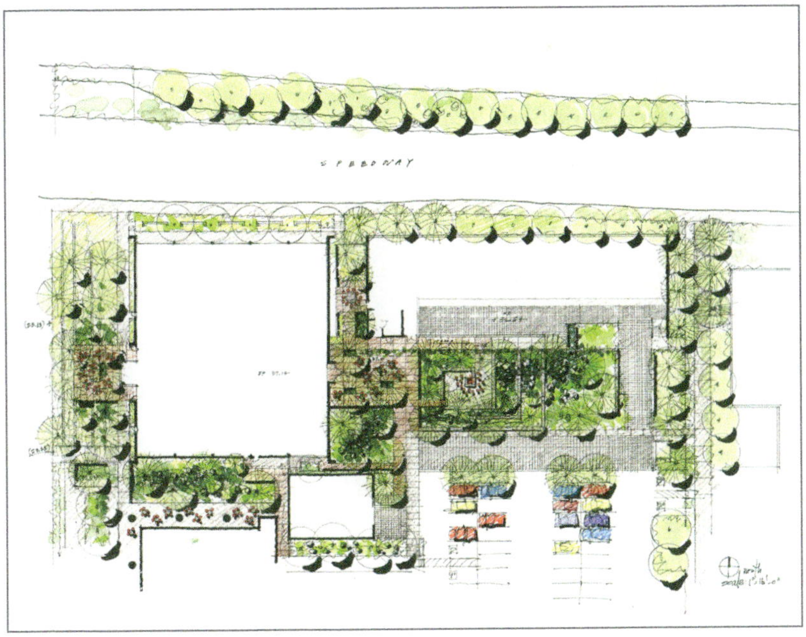

TEN EYCK LANDSCAPE ARCHITECTS

FIGURE 1.5 Site plan.

THE SITE: Asphalt campus parking lot located adjacent to the School of Architecture and Landscape Architecture at the University of Arizona. The Tucson climate is hot during the summer and cool in winter. Average annual precipitation is 12 inches.

continues

UNDERWOOD FAMILY SONORAN LANDSCAPE LABORATORY *(CONTINUED)*

Design Overview

In 2006, the University of Arizona built a new expansion facility that brings students from architecture, planning, and landscape architecture under one roof to provide an integrated learning environment. The asphalt parking lot adjacent to the school was transformed into the Underwood Family Sonoran Landscape Laboratory, which functions as both an outdoor classroom and entry plaza. The research-oriented garden serves as a demonstration facility that focuses on water-conscious design solutions and functions as a cleansing biosponge for stormwater runoff and building wastewater (see Figure 1.5).

Five distinct ecological communities of the Sonoran Desert are represented in the desert laboratory. The 5,000-gallon (18,900 L) pond provides habitat for endangered fish and is listed by the U.S. Fish and Wildlife Service as a "safe harbor" urban site (see Figure 1.6). The diverse garden is vegetated with native drought-resistant plants appropriate for each biome. A vertical scrim extends along the south side of the building and is vegetated with vines that have climbed 50 feet (15.24 m) high, which help to reduce the building's energy consumption.

BILL TIMMERMAN

■ **FIGURE 1.6**
Wetland pond and shaded lower court. The 5,000-gallon (18,900 L) pond provides habitat for endangered fish and is listed by the U.S. Fish and Wildlife Service as a "safe harbor" urban site.

UNDERWOOD FAMILY SONORAN LANDSCAPE LABORATORY *(CONTINUED)*

Extensive collaboration between the project architect, landscape architect, engineers, and irrigation consultant resulted in an impressive water harvesting system that collects rainwater from the roof, air-conditioning condensate, and greywater from the building's drinking fountains. The water is stored in an 11,600-gallon (43,911 L) cistern and over the course of a year, 244,000 gallons (922,320 L) are harvested. The recycled water is comprised of approximately 40 percent condensate, 33 percent rainwater runoff, 18 percent well water blowoff and 9 percent greywater. The well's operation requires daily flushing, which was sending 200 gallons (757 L) per day to the city storm drain system. The fresh water from the blowoff is now diverted into the desert riparian pond and helps to maintain water levels and the appropriate conditions for the desert fish species. After the initial establishment period, the site's water use will be balanced; potable water will likely no longer be required, and the garden will rely solely on reclaimed water sources (see Figure 1.7).

BILL TIMMERMAN

■ **FIGURE 1.7**
Native plants adapted to the site conditions are planted throughout the site. Once the vegetation is established, potable water will likely no longer be required, and the garden will rely solely on reclaimed water sources.

continues

UNDERWOOD FAMILY SONORAN LANDSCAPE LABORATORY *(CONTINUED)*

PROJECT TEAM

- **LANDSCAPE ARCHITECTS**
  Christine E. Ten Eyck, FASLA
  Todd Briggs, ASLA, project manager
  www.teneyckla.com
- **ARCHITECT**
  Jones Studio
  www.jonesstudioinc.com
- **CIVIL ENGINEER**
  Evans Kuhn
  www.evanskuhn.com
- **MECHANICAL ENGINEER**
  Kunka Engineering
  www.kunka.com
- **IRRIGATION DESIGN**
  Carl Kominsky
- **WETLAND CONSULTANT**
  Wass Gerke & Associates
  www.azwetlands.com
- **GENERAL CONTRACTOR**
  Lloyd Construction Company, Inc.
  www.lloydconstruction.com
- **LANDSCAPE CONTRACTOR**
  AAA Landscape
  www.aaalandscape.com

Creating a Love for Nature in Our Children

Children who feel connected to the natural environment and the ecological processes that sustain humanity are better equipped to face the challenge of building a sustainable society. Unfortunately, today's children are spending less and less time outdoors and as a result, their knowledge and appreciation of the natural world is dwindling (Louv 2005). The increasing disconnect with nature can be attributed, in part, to residential and schoolyard landscapes that children often find boring and uninspiring and to the layout of our neighborhoods and communities, which often limits safe access to natural settings (Moore and Marcus 2008).

Children are fascinated by nature and have an innate desire to splash in water, chase butterflies, get muddy, and explore their surroundings (see Figure 1.8). If their curiosity is not given an opportunity to flourish, an aversion to nature—or biophobia—may develop, which can result in a general discomfort, fear, or disregard for the natural environment (Kellert and Wilson 1993).

■ **FIGURE 1.8** Children playing with rocks and water that are part of a cleansing biotope at Tanner Springs Park.

In order to cultivate a love for nature within children, they must first have fun playing outdoors and immersing themselves in healthy ecosystems and all of their components. Providing these opportunities where children spend their days—at home or school, or in a local park—enables spontaneous interaction with nature to become part of everyday life and relieves parents of the need to program time in the natural world into children's lives (Moore and Marcus 2008). Unstructured, child-directed play in "wild" settings—as opposed to structured or programmed activities such as planting a tree or caring for a plant—has been found to be more effective at encouraging developmental impacts that support an environmental ethic in adults (Wells and Lekies 2006). Sites can serve a special and valuable purpose when they encourage children to play outdoors and explore the natural environment (see Figure 1.9).

■ **FIGURE 1.9** Fifty-foot-long hillside slide integrated into the Adventure Garden at the San Francisco School. The terraced garden is built from recycled concrete taken from a demolished basketball court located on-site. The schoolyard integrates concepts of sustainability, recycling, and reuse into the physical form of the landscape.

Continual Improvement: Monitoring and Adaptive Management

The living systems that make up a sustainable site do not exist in a fixed state. Similar to natural ecosystems, they grow, senesce, and evolve over time. The same is true for the culture of a site and how people choose to use and experience the landscape. Acknowledging that change is an unavoidable and essential component of a site is key to the long-term success of the project.

Postoccupancy evaluations and the monitoring of sustainable design practices are necessary for continued improvement and informed site stewardship. Adaptive management uses the information gathered to continually adjust maintenance practices and improve the overall function of the site.

Planning for information gathering and adaptive management begins in the design phase. Project teams can incorporate tracking mechanisms into the site design for water and energy use, waste disposal, and other performance targets. And the design of the site can ease the gathering of information and encourage monitoring.

To understand which components of a site to monitor, the goals and performance targets of the project must first be agreed upon. How monitoring will be used to improve site performance should be clear to all those involved. Projects are more likely to be successful when the individuals collecting and using the data are included in the design process.

Project teams may need to educate clients about the public perception and monetary benefits of monitoring and adaptive management, which include:

Avoiding trial and error maintenance practices

Reducing replacement costs

Preventing extreme overhauls of failing systems

In addition to the on-site benefits, postoccupancy evaluations and monitoring also provide invaluable opportunities for continued learning that can improve the body of knowledge and success of the sustainable design industry.

■ **TABLE 1.2** Guiding principles are commonly held values or fundamental beliefs that steer an organization, team, or individual's decision making. They are the foundation of the design process and help articulate expectations and evaluate success.

GUIDING PRINCIPLES OF A SUSTAINABLE SITE
DO NO HARM. Avoid making changes to the site that will degrade the surrounding environment. Promote projects on sites where previous disturbance or development presents an opportunity to regenerate ecosystem services through sustainable design.
OBSERVE THE PRECAUTIONARY PRINCIPLE. Be cautious in making decisions that could create risk to human and environmental health. Some actions can cause irreversible damage. Examine a full range of alternatives—including no action—and be open to contributions from all affected parties.
DESIGN WITH NATURE AND CULTURE. Create and implement designs that are responsive to economic, environmental, and cultural conditions.
PROVIDE REGENERATIVE SYSTEMS AS INTERGENERATIONAL EQUITY. Provide future generations with a sustainable environment supported by regenerative systems and endowed with regenerative resources.
SUPPORT A LIVING PROCESS. Continuously reevaluate assumptions and values and adapt to demographic and environmental change.
USE A SYSTEMS-THINKING APPROACH. Understand and value the relationships in an ecosystem and use an approach that reflects and sustains ecosystem services; reestablish the integral and essential relationship between natural processes and human activity.
USE A COLLABORATIVE AND ETHICAL APPROACH. Encourage direct and open communication among colleagues, clients, manufacturers, and users to link long-term sustainability with ethical responsibility.
CONTINUALLY IMPROVE SITE PRACTICES. Conduct postoccupancy evaluations and ecological monitoring to inform the maintenance of the site and provide opportunities for continued learning that improves the field of sustainable design.
FOSTER ENVIRONMENTAL STEWARDSHIP. In all aspects of land development and management, foster an ethic of environmental stewardship—an understanding that responsible management of healthy ecosystems improves the quality of life for present and future generations.
CONNECT PEOPLE TO NATURE. Create environments where all people can receive and enjoy the benefits of nature in their everyday lives.

SOURCE: THE SUSTAINABLE SITES INITIATIVE GUIDELINES AND PERFORMANCE BENCHMARKS, 2009

■ CASE STUDY

PACIFIC CANNERY LOFTS

PROJECT TYPE: Mixed-use, multifamily development

LOCATION: Oakland, California

SIZE: 2.7 acres (1.1 hectares)

COMPLETION DATE: 2008

CLIENT: Holliday Development

HIGHLIGHTED SUSTAINABLE PRACTICES:

- Redevelopment of a brownfield site
- Within walking distance to mass public transportation
- Reuse of existing on-site materials
- Reduces impervious cover
- Increases vegetative biomass
- Gardens include edible plants
- Mitigates the urban heat island
- Utilizes reclaimed water in a drip irrigation system

THE SITE: Industrial brownfield site located in West Oakland. The historic neighborhood was characterized by abandoned warehouses, a crumbling train station, and a maze of raised freeways, frontage roads, and rail lines.

■ **FIGURE 1.10** Pacific Cannery Lofts master plan.

Design Overview

Pacific Cannery Lofts is an adaptive reuse project that has transformed a historic vegetable cannery into 163 contemporary loft and town house units (see Figure 1.11). The site is part of a vision to redevelop nearly 30 acres of brownfield into a new Central Station neighborhood that brings together a number of developers to build parks with improved streets, commercial spaces, an urban farm, and over 1,000 new housing units around the renovated train station.

The site design features three internal garden courtyards that are linked by a 350-foot-long (107 m) double-height corridor known as the Gallery, which serves as the internal "main street" of the project. A sense of retreat and privacy for residents was created through a thoughtful organization of space, rich detail in the lushly planted courtyards, and a linear grove court featuring fruit trees and edible plants. Central walkways focus circulation to the middle of the spaces, leaving room adjacent to the buildings for individual entry garden zones and privacy plantings designed to screen private unit patio areas (see Figure 1.11). Incorporating edible plants and highlighting natural wind and stormwater events in the gardens tempers the heavily built atmosphere of the site.

continues

PACIFIC CANNERY LOFTS *(CONTINUED)*

The main entry courtyard is designed as a rain garden. Flagstone paths lead to individual unit entries furnished with a dual-purpose bench and aqueduct. Water cascades from the pebble-filled aqueduct into linear "rivers" adjacent to the main walkway that hold and cleanse the water before it infiltrates into the local aquifer. Recycled tumbled glass installed at the surface of the channels is underlit with LED strands marking the path and giving the courtyard a warm glow at night (see Figure 1.12). Reclaimed gears and valve heads embedded in the walkway provide rhythm and indicate locations of drain inlets set immediately below the recycled tumbled glass. The drain inlets relieve the courtyard when extreme downpours deluge the infiltration system, thereby protecting the building from flooding.

Abandoned cannery relics are reused throughout the Pacific Cannery Lofts project as industrial sculpture. Ten-foot diameter cast-iron wheels, originally part of the cannery's ice-making equipment, mark the west entry, along with engines, mounts, and other related machine parts. The cannery's original scale marks the east entry, and slate-plated switching stations are set in the building's new gallery arcade.

■ **FIGURE 1.11**
Central walkways in the dining room courtyard focus circulation to the middle of the space, leaving room adjacent to the buildings for individual entry garden zones and privacy plantings designed to screen private unit patio areas.

■ **FIGURE 1.12**
Flagstone paths lead to individual unit entries furnished with a dual-purpose bench and aqueduct. Water cascades from the pebble-filled aqueduct into linear "rivers" adjacent to the main walkway that hold and cleanse the water before it infiltrates into the local aquifer. Recycled tumbled glass installed at the surface of the channels is underlit with LED strands marking the path and giving the courtyard a warm glow at night.

PACIFIC CANNERY LOFTS *(CONTINUED)*

PROJECT TEAM

- **LANDSCAPE ARCHITECTS**

  Miller Company Landscape Architects

  www.millercomp.com

  Jeffrey Miller, Principal Landscape Architect

  Leah Hickey, project assistant

- **ARCHITECTS**

  David Baker + Partners Architects

  www.dbarchitect.com

- **GENERAL CONTRACTOR**

  Cannon Constructors

  www.cannongroup.com

- **LANDSCAPE CONTRACTOR**

  Miller Company Landscape Contractors

  William Rogers, project manager

  www.millercomp.com

■ **FIGURE 1.13**
Brightly hued custom concrete banquettes and low tables flank the central walkway in the living room courtyard. The tandem U-shaped seating design invites conversation and provides respite. Large leaf and flower plantings create a tropical effect, while the low-water-use understory provides texture and fragrance.

REFERENCES

American Farmland Trust. 2009. Farming on the edge report. http://www.farmland.org/resources/fote/default.asp; accessed 10/11/11.

Brundtland, G. H. 1987. *Our common future: Report of the World Commission on Environment and Development*. Oxford University Press.

Costanza, R., et al. 1997. The value of the world's ecosystem services and natural capital. *Nature* 387 (May 15): 253–60.

Kellert, S., and E. O. Wilson. (1993). *The biophilia hypothesis.* Washington, DC: Island Press.

Leopold, A. 1949. *A Sand County almanac: And sketches here and there.* Oxford University Press.

Meyer, E. K. 2008. Sustaining beauty: The performance of appearance. *Journal of Landscape Architecture* (Spring).

Moore, R.C., and C. C. Marcus. (2008). Healthy planet, healthy children: designing nature into the daily spaces of childhood. In *Biophilic design,* by S. Kellert, J. Heerwagen, and M. Mador. Hoboken, NJ: John Wiley & Sons, pp. 153–204.

Wells, N. M., and S. K. Lekies. 2006. Nature and life course pathways to environmentalism. *Children, Youth and Environments* 16:1–24.

World Wildlife Fund. 2010. Living planet report 2010: Biodiversity, biocapacity and development.

■ **FIGURE 2.1**
Turtle Creek Pump House.

CHAPTER 2

The Sustainable Site Design Process

SUSTAINABLE SITE DESIGN is a creative and analytical process of information gathering, investigation, and composition that utilizes art and science to connect natural and built systems in a mutually beneficial way. Design outcomes are not inherently sustainable and should not be assumed just because a site is made up of vegetation, soil, and other natural components. Like all successful aspects of a project, sustainability must be intentional and nurtured.

Project teams should view each design decision as an opportunity to reduce consumption, eliminate waste, cultivate healthy ecosystems, and connect people with nature. Beneficial impacts are limited when sustainable design is considered separately from the overall design process and reduced to intermittent "green" components—such as native plants or recycled materials—that are tacked on to a project. Such a piecemeal approach paints sustainability as a nice but unnecessary luxury that is inessential to the project. By infusing sustainability into all aspects of the design, it becomes an interwoven and inseparable component that is vital to the project's overall success.

Traditional design processes and team interactions do not always support sustainable outcomes. To help overcome this issue, designers must use an integrated design process, in which teams work together in a collaborative fashion and utilize the technical expertise of other professions to broaden their awareness of the range of possible design solutions.

Integrated Design and the Multidisciplinary Design Team

Multidisciplinary Team

Sustainable site development requires holistic thinking and a wide spectrum of expertise and skills best obtained through a multidisciplinary and integrated design effort. The design team should, at a minimum, include the client and professionals proficient in the local ecology as well as sustainable landscape design, construction, and maintenance practices. Depending on the particular criteria of the site and design program, additional expertise may be required. In some circumstances, one person may play multiple roles: a homeowner who is the client, for example, may have the experience and interest in maintaining his or her landscape. Situations in which one person serving multiple roles may impact quality control measures or generate conflicts of interest should be avoided.

Integrated Design

Integrated design is an iterative process of research and analysis, communication, and design exploration that occurs collectively among all team members throughout all phases of the project (7group and Reed 2009). Whereas the conventional design process is typically a linear approach, comprised of a collection of discrete tasks that often proceed from owner to landscape architect to subconsultants to general contractor to subcontractor to site user, the integrated process encourages the multidisciplinary team to be collectively involved in the design process and utilizes their various perspectives to develop design solutions holistically. The process recognizes the relationship among the site's physical, biological, and cultural components, and, because of this recognition, is aware of a more complete set of design opportunities and impacts (Keeler and Burke 2009).

To establish a culture of design integration, projects often begin with a charrette or other collaborative setting that creatively explores design options, uncovers areas of conflict, and establishes the project concept. Team members are expected to provide input and discuss areas beyond their conventional areas of expertise to help reveal how their work will interact with and affect other portions of the project. Bringing the multidisciplinary team together to explore the site's environmental and social systems encourages synergy and the optimization of design solutions early in the process, thus limiting environmental impacts and saving time and money over the life of the project (Mendler, Odell, and Lazarus 2006). Throughout the design process, the multidisciplinary team repeatedly assembles to share research and analysis findings, discuss options, and discover new opportunities. With a collaborative focus, the team then separates to design and analyze, with the intent of reassembling at the next juncture (7group and Reed 2009). This open dialogue builds trust and mutually supportive working relationships among team members.

In circumstances in which the integrated design process requires higher design fees, the diverse problem-solving approach often leads to lower construction and reduced maintenance costs (Keeler and Burke 2009). For example, including the land-care professional in the design process provides opportunities for discussing the maintenance requirements of the proposed design solutions and adjusting them accordingly to eliminate unnecessary site damage or design solutions that may require expensive and timely maintenance. Input from the contractor may result in creative opportunities to reuse existing site structures and materials that reduce waste and speed the construction process. Guidance from an ecologist can provide strategies for protecting healthy soil and vegetation during construction, improving site performance and avoiding restoration and replacement costs.

The key to achieving an integrated design is to maintain coordination and collaboration among team members throughout all project phases. Even though the integrated design process actively seeks input of all team members, it is not design by committee (Yudelson 2009); a project manager involved in all aspects of the project is still necessary. However, it is important that the project manager act more as a team leader and genuinely welcome input from all team members, rather than as a sole decision maker. Giving all members of the team an opportunity to engage in the design process and vet concerns results in a significantly higher level of project ownership and a commitment to achieving the project's goals and performance targets within the project budget (7group and Reed 2009).

Strategies for encouraging integrated design and multidisciplinary collaboration include

- Developing consensus on the strategies and tools that will be used to share information and foster collaboration
- Structuring the project schedule to allow time for integrated design and reflection
- Clarifying and communicating the roles and responsibilities of each team member
- Diagramming the design process and creating feedback loops for each phase, noting where professionals will be collaborating and why
- Developing consensus on the project goals, sustainable guidelines, and performance targets
- Focusing charrettes and other collaborative design activities at the beginning of the design process; encouraging team members to explore design solutions from multiple perspectives and utilizing team knowledge to create innovative solutions
- Conducting regular team meetings scheduled around project milestones, mandating active participation from all team members, and looking for overlapping benefits and opportunities across disciplines
- Encouraging interim meetings between team members to continue information sharing and collaboration
- Sincerely soliciting and integrating the input of other team members; viewing the diversity of opinions as an asset and using it to thoroughly analyze and explore design solutions

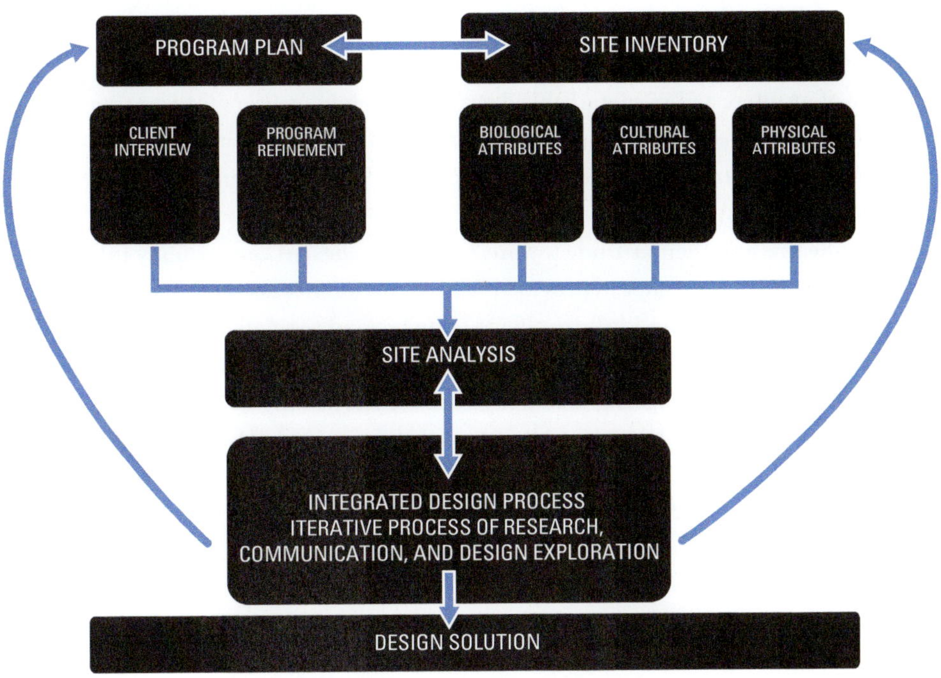

FIGURE 2.2 The integrated design process is an iterative process of research, communication, and design exploration. Integrated design brings the project team together at important junctures to share research and analysis findings, discuss options, uncover new opportunities, and make design decisions. Throughout the process, additional site inventory and analysis is often required, and the program plan may need to be revisited to accommodate new challenges or criteria that are revealed.

Assembling the Design Team

The success of a project is largely dependent on the ability and commitment of the design team. Ideally, the team would be made up of professionals who specialize in sustainable solutions and have valuable project experience. When this is not the case, sustainable outcomes can still be successfully achieved if the project team is carefully assembled and expectations are made clear (Mendler, Odell, and Lazarus 2006).

At a minimum, all team members should be competent professionals who meet the following criteria:*

- They are open to new ideas.
- They have a positive attitude toward developing an innovative project.
- They are comfortable with an integrative design process that questions conventional assumptions and tests new ideas.
- They are committed to going beyond minimum code performance and achieving sustainable outcomes.
- They are willing to work collaboratively, navigate obstacles, and learn from others.

Team members who do not sincerely meet this criteria can stifle progress, prevent innovation, and increase overall project time and costs. Design teams with centrally aligned goals and expectations are more likely to create enjoyable and beneficial professional experiences for all those involved and are better suited to creating a successful project within budget parameters.

Defining the Project

Inspiring the Client

Clients and citizens are becoming increasingly aware of the benefits of green building and as a result are incorporating sustainability into their definition of project success. According to a survey of 381 firms conducted by the American Society of Landscape Architects, 96 percent of clients were knowledgeable of or interested in sustainable design. The driving factors for wanting to incorporate sustainability into projects were reduced utility and maintenance costs, government regulation, code or construction standards, marketing cachet, and reducing environmental harm (American Society of Landscape Architects 2009).

Design teams and their attitude toward sustainable design can strongly influence the client's desires and expectations. Teams should provide the information necessary to help clients feel comfortable with green building practices and inspire clients to go beyond regulated standards. Initial client meetings are an opportunity to gain an understanding of the client's sustainable design knowledge and interest as well as beginning an ongoing dialogue about the environmental and health benefits landscapes can provide to site users and the surrounding region. Clients should be engaged in the design process and brainstorming of sustainable design solutions. In many instances, it can be helpful to visit or provide examples of other sustainable projects.

* Mendler, Odell, and Lazarus 2006; Kwok and Grondzik 2007.

Acknowledging a Commitment to Sustainability

As part of the project contract, the client and design team typically agree on the general scope of the project and outline the services to be provided. Traditionally, these documents have not included an integrated design process or the necessary steps to achieve sustainable outcomes (Mendler, Odell, and Lazarus 2006).

In order to clarify expectations from the start, the project contract must acknowledge a commitment to sustainability. Contracts should clearly document expectations that team members will use an integrated design process and actively participate in research and analysis, project meetings, and collaborative design exploration. The role of team members in helping establish, track, and achieve the sustainable guidelines and target performance benchmarks should be carefully outlined. Where applicable any green certification expectations or desires, such as the Sustainable Sites Initiative (SITES), Leadership in Energy and Environmental Design (LEED), or the British Research Environment Environmental Assessment Method (BREEAM) should be acknowledged and identified.

It may be helpful to develop team member contracts in stages, with the first agreement focusing on the time required to prepare and participate in the initial meetings and design charrette. The final contract can then be developed based on the outcomes of the charrette and established design direction. This approach creates project scopes more accurately and avoids overpricing stemming from unknown circumstances (Mendler, Odell, and Lazarus 2006).

Initial Team Meeting

The integrated design process requires individual professionals to come together as a team and support one another in achieving the project's goals and performance targets. This collaborative effort can be challenging, and a diversity of values, opinions, expectations, and perspectives are to be expected. Diversity should be viewed as an asset and used to thoroughly analyze and explore design solutions. When options are openly discussed, a cross-pollination of design concepts is more likely to occur and result in hybridized solutions that represent the best ideas from each profession.

Prior to beginning design, take time to develop a cohesive team and establish a culture of communication and collaboration. Discussing the following items will aid the team building process and lay the groundwork for a successful project.

- Briefly review the guiding principles or core values that direct the team's work.
- Introduce the team to the fundamentals of the integrative design process.
- Develop consensus on the strategies that will be used to foster collaboration.
- Discuss the preliminary project schedule and the expectations of team members.
- Address any concerns team members may have with the project (e.g., permitting, liability, scheduling) so that the issues can be addressed and overcome.

Understanding the Site

Site Assessment: Inventory and Analysis

SITE INVENTORY

A broad knowledge and thorough understanding of the local ecology and culture is essential to the design and development of a sustainable site. Each site has a unique set of physical, biological, and cultural attributes that define the overall character of the landscape and determine its suitability for specific uses (LaGro 2008). All too often design teams proceed without a full understanding of the living systems and communities they are impacting. When the context of a site is not well understood, design decisions can unnecessarily and unknowingly lead to damaging environmental, social, and economic outcomes.

Site inventories communicate and map the physical, biological, and cultural components of a site and the surrounding area. This initial reconnaissance provides the information required to begin the design process; as the project develops, additional information is gathered to inform design solutions. The inventory is not intended to be an open-ended process of information gathering but rather a focused compilation of site conditions prompted by the requirements of the program plan and questions or concepts that arise during the design process.

Understanding the full context of a site may require multiple visits from a variety of specialists. Expertise from ecologists, hydrologists, soil scientists, or building engineers may be necessary to collect, map, and analyze the information needed to fully and accurately appreciate the opportunities and limitations of the site. Including specialists in the assessment aids the design team in understanding the current and potential function of the site's systems and provides opportunities to optimize design solutions within the existing project parameters (see Table. 2.1).

■ **TABLE 2.1** Site Inventory

REGIONAL CONTEXT

SITE INVENTORY	INFORMATION GATHERING	DESIGN CONSIDERATIONS
Identify regional environmental and human health issues or concerns such as air pollution, combined sewer overflow, or water shortages.	Contact local and regional health and environmental authorities, regional planners, and community leaders.	Sites are part of a larger ecological and social community. Sustainable landscapes not only prevent environmental damage but also remedy existing problems at the site scale and beyond. It is important to understand the surrounding conditions and explore design options that mutually benefit the site and surrounding area. Design teams should weigh design options and make decisions based on the solutions that will provide the greatest benefits. Developing connections to the community and supporting the local character of a region enhances feelings of stewardship and sense of place. Understanding the local context also allows the project team to identify and mitigate any negative impacts from surrounding sites.
Identify the eco-region and the area's major native plant communities and environmental conditions.	Research eco-region maps from the U.S. EPA or similar organization. Field-check and compare descriptions to the actual site conditions.	
Research existing comprehensive community plans and zoning codes that may influence the site.	Contact local planning agencies and authorities.	
Study the surrounding area and identify adjacent site conditions and current uses. Determine whether the surrounding conditions will benefit the site or have a negative impact. Note any aesthetically pleasing visual qualities and stressful factors, such as excessive noise, odor, or pollution.	Explore the area surrounding the site to become familiar with the local culture, amenities, and community resources. Interview neighbors, community leaders, and other project stakeholders. Utilize tools such as Google Earth and GIS.	
Identify areas of wildlife habitat and migratory routes and corridors in the areas surrounding the site.	Contact local wildlife authorities. Conduct a regional habitat inventory. Interview neighbors, community leaders, and other project stakeholders.	

■ **TABLE 2.1** Site Inventory *(continued)*

SITE INVENTORY	INFORMATION GATHERING	DESIGN CONSIDERATIONS
Identify potential for damage to the site from natural disasters such as hurricanes, wildfire, and floods.	Research the natural disaster history of the area. Interview community residents and local authorities.	Thoughtful site selection, design, and management can reduce the risk and impact of natural disasters. Special attention should be given to building location, materials, and construction methods.
Identify existing and planned public transit and bicycle or pedestrian systems located within ¼ mile of the site.	Contact local and state transportation authorities.	Mass transit and other alternative transportation options reduce the generation of greenhouse gases and improve air and water quality. Understanding the local transportation systems provides opportunities for the site to connect with—and encourage the use of—public transit and nonmotorized transportation.

CLIMATE AND ENERGY

SITE INVENTORY	INFORMATION GATHERING	DESIGN CONSIDERATIONS
Determine the average precipitation, humidity, and temperature of the site for each month of the year.	Research historical weather data from local meteorologists, weather stations, and universities.	Sites that are designed to thrive in their natural climatic conditions require fewer resources to sustain. Rainfall and temperatures affect design issues such as vegetation and material selection, stormwater management, and site layout.
Identify on-site conditions that provide opportunities for renewable energy strategies such as wind, solar, and geothermal.	Contact state and local energy authorities.	Renewable energy sources reduce greenhouse gas emissions and air pollution from fossil fuels. Consider the effects of existing vegetation, topography, and structures that may cast shadows or act as windbreaks.

MICROCLIMATE

SITE INVENTORY	INFORMATION GATHERING	DESIGN CONSIDERATIONS
Study the path of the sun. Determine shadow configurations from trees, topography, and structures.	Create a solar path diagram for the site. Map the path of the sun through the day and year.	Sites often have unique microclimatic conditions that differ from regional weather patterns. Understanding the microclimate allows the design team to utilize and create site conditions that increase user comfort and reduce buildings' energy consumption. Special attention should be given to building orientation and plant and material selection.
Research ground-level prevailing wind direction in all seasons. Consider the effects of site features such as topography, vegetation, and buildings.	Create or study existing wind rose diagrams. Research historical weather data from local meteorologists, weather stations, and universities.	
Identify surfaces that heat or cool the site, such as bodies of water or dark pavements or roofs.	Field-check locations and surface materials. Cross-reference findings with wind direction and shadow patterns to determine the effects on the microclimate.	

continues

■ **TABLE 2.1** Site Inventory *(continued)*

HYDROLOGY		
SITE INVENTORY	**INFORMATION GATHERING**	**DESIGN CONSIDERATIONS**
Study the site topography. Map the natural flow of water and areas of ponding.	Study topographic maps derived from satellite imagery or physical surveys. The topographic detail required will depend on the size of the site and specific design objectives.	Topography influences many aspects of the site, such as the microclimate, distribution of plant and animal species, water movement, and soil depth. Consider options for minimizing disturbance and artfully incorporating the existing topography into the design solution.
Estimate the volume of rainwater or other nonpotable water sources such as stormwater, greywater, and air-conditioner condensate available on-site for reuse.	Work with building architects and engineers to understand the building's water use and wastewater flow. Utility bills and other building records can be useful in establishing baseline data.	Sustainable sites treat all water as a resource and strive to promote water quality and support healthy hydrologic processes. Potable water requirements can be reduced or eliminated through design strategies such as rainwater harvesting or greywater and air-conditioner condensate reuse.
Map the one-hundred-year floodplain.	Consult federal flood maps, state environmental agencies, or local studies to determine the one-hundred-year floodplain.	Development of floodplains or alterations in floodplain topography can increase the risk of flooding, water pollution, and property damage both on-site and downstream.
Map existing water bodies (e.g., lakes and streams) and their associated shorelines or vegetated buffer zones. Describe existing conditions, such as habitat quality, bank stability, and any artificial modifications. Note ecological restoration opportunities.	Use aerial photos or site maps to locate existing water bodies. Ground truth the location and extent of vegetated buffer. Locate healthy habitats within the region that can be used as reference sites.	Changes to water bodies and their associated buffers are often regulated by state and federal authorities. Project teams should consider the impact of site design, construction, and maintenance decisions on the quality, habitat, aesthetic, and recreational value of the water bodies.
Locate and delineate existing wetlands and their associated buffers.	Conduct soil and plant surveys to identify wetland areas.	Wetlands are protected by state and federal authorities. Altering the site's drainage patterns, soil conditions, and groundwater levels can impact the health of wetlands. Existing natural wetlands should not be used for stormwater management or wastewater practices.
Identify water bodies on-site or downstream from the site that are listed as impaired by the state water quality agency. Determine the specific pollutants of concern for the impaired water bodies.	Research the federal and state water quality data. In the United States, see the Clean Water Act Section 303(d) list provided by the state water quality agency.	Through careful design and maintenance, sites can reduce pollutant sources and the volume of stormwater runoff. Pay special attention to the selection of materials, on-site treatment of stormwater, and maintenance practices that minimize pollutant loads.
Identify sources or potential sources of water pollution and health hazards existing on-site.	Research existing drainage infrastructure. Identify the water source, treatment location, and strategy. Field-check and identify building, hardscape, and landscape materials such as treated lumber or galvanized metal that might be sources of pollutants. Interview the maintenance contractor or other individuals responsible for the site's care to identify potential pollution sources.	Building materials and maintenance practices can be pollution sources. Water quality impacts should be considered when selecting materials and construction and maintenance strategies.
Determine seasonal groundwater elevations	Landscape cues such as springs, seeps, and water-loving vegetation can indicate areas of shallow groundwater. Use groundwater monitoring wells or similar technology to accurately determine the groundwater depth across the site.	Groundwater elevations can impact the site's hydrology and suitability for excavation, stormwater and wastewater management, and other site features. Special consideration should be given to site development and maintenance strategies to avoid the contamination of groundwater.

■ **TABLE 2.1** Site Inventory *(continued)*

SOILS

SITE INVENTORY	INFORMATION GATHERING	DESIGN CONSIDERATIONS
Research the site geology and subsoil conditions.	Obtain soil survey maps from federal and local agencies. Conduct field tests to verify conditions.	Geology influences a site's suitability for excavation, grading, wastewater disposal, stormwater management, pond construction, and other landscape amenities.
Determine the soil type(s) and document characteristics such as pH, permeability, erosion potential, and depth. Field-check and map healthy and degraded soil conditions. Determine the areas to be protected and those best suited for development. Soil conditions can change over short distances; therefore, it cannot be assumed that soil health is consistent across the site.	Obtain soil survey maps from federal and local agencies. Conduct soil surveys and field tests to verify conditions. Design teams can look for landscape changes such as variances in soil color or vegetative cover that may indicate different soil conditions.	Healthy soils provide a variety of ecosystem services such as water cleansing and storage, carbon sequestration, and habitat. Protecting healthy soils reduces restoration costs and improves plant performance. Areas of degraded soils should be considered first for design elements that require significant soil and vegetation disturbance.
Investigate the site to determine if any soils are categorized as prime farmland, unique farmland, or farmland of statewide importance by the National Resources Conservation Service (NRCS).	In the United States, obtain NRCS soil maps of the site. In areas where maps are not available, contact the local NRCS office for more information.	Prime farmland, unique soils, and soils of statewide importance produce crops more efficiently than other soils, requiring fewer inputs, such as fuel, water, and fertilizers. The development of these unique and high-quality soils should be avoided.

VEGETATION

SITE INVENTORY	INFORMATION GATHERING	DESIGN CONSIDERATIONS
Identify and map vegetative communities (i.e., woodland, tall grass prairie, riparian). Conduct a qualitative inventory of the community. Note wildlife species associated with the community. Record current maintenance and management practices. Identify areas to be protected or those suitable for development.	Conduct plant and wildlife surveys. Review aerial photos and satellite imagery.	Both urban and rural sites can provide a variety of plant and animal habitats. Design teams should look for opportunities to protect and restore habitat on-site as well as connect with surrounding areas of habitat. The development of threatened or endangered species habitat should be avoided. Areas of low-quality habitat should be considered first for design elements that require significant soil and vegetation disturbance.
Investigate the site for habitat that may support threatened or endangered plant and animal species.	Research federal and state threatened or endangered species lists. Contact local state agencies for guidelines on conducting species surveys and development requirements.	
Survey existing site vegetation. Create a vegetative cover map that identifies (1) trees over 6 inches diameter at breast height or as required by local ordinance, (2) heritage or special-status trees, (3) invasive species, and (4) other significant or dominant vegetation. Generate a general species list of dominant vegetation in the canopy, subcanopy, and herbaceous layers. Include common and Latin names. Estimate the frequency or percent cover. Note any unusual or unique vegetation. Determine whether the vegetation is native to the region.	Conduct vegetative surveys. Review aerial photos and satellite imagery.	Incorporating existing vegetation into the site design provides a variety of environmental and economic benefits. Design teams should look for opportunities to minimize disturbance and artfully incorporate existing vegetation into the design solution. Areas including invasive species or unhealthy or other undesirable vegetation should be considered first for design elements that require significant soil and vegetation disturbance.

continues

■ **TABLE 2.1** Site Inventory *(continued)*

MATERIALS

SITE INVENTORY	INFORMATION GATHERING	DESIGN CONSIDERATIONS
Identify and map existing buildings and landscape materials, such as outdoor structures, roads, and pathways. Note the materials' size, condition, and potential for reuse or recycling.	Review site surveys and aerial and satellite imagery. Field-check all surfaces and materials to determine conditions.	Reusing and recycling materials reduces the use of virgin feedstock, which in turn minimizes habitat destruction, waste generation, and air and water pollution. Throughout the design and construction processes, explore opportunities to reuse or recycle on-site materials.

CULTURAL INVENTORY

SITE INVENTORY	INFORMATION GATHERING	DESIGN CONSIDERATIONS
Research the site's history and prior uses.	Interview property owners and neighbors. Research city and county records and historic aerial photos. Survey the site for indicators of previous use. Test for possible contamination.	Understanding the site history and prior use is helpful in identifying conditions that may not be obvious or are unforeseen, such as contamination or soil instability.
Determine the locations of existing public infrastructure, such as roads and utility networks.	Contact local utility and transportation agencies and authorities.	Existing public infrastructure can influence the placement of items, such as buildings and site entrance and egress locations.
Identify project stakeholders.	Work with neighborhood leaders and other local "experts" to identify individuals and groups that need to be involved in the design process.	Site users and other stakeholders can provide unique insight and become active stewards of the site.
Document existing site uses and their associated user groups.	Observe the site during various times of the day. Interview site stakeholders.	Locations with a unique sense of place connect the community to the site and encourage stewardship. Look for opportunities to get project stakeholders involved in the design process and provide feedback on the site conditions and amenities they value.
Identify and map historical or cultural landscape features. Map characteristic site features that are unique or memorable, such as rock outcroppings or view corridors.	Observe the site during various times of the day. Interview neighbors, community leaders, and other project stakeholders. Contact historical commissions and associations.	
Identify potential or existing odors, noise pollution, or unsightly features that may be considered an annoyance.	Observe the site during various times of the day. Interview site users and other project stakeholders. Determine the source of odor and direction of prevailing winds. Use a sound-level meter to measure the level of noise. Follow ASTM E1014-08 Standard Guide for Measurement of Outdoor A-Weighted Sound Levels or similar standard noise measurements.	Existing landscape features that are loud or unsightly can have negative impacts on the site users' experience. Give special attention to the location of existing and planned equipment, such as heating, ventilating, and air-conditioning (HVAC) systems. Locate design components such as buildings or vegetation to screen or block unwanted views and sounds.

Site Analysis

Site analyses interpret inventory information to identify the areas best suited for specific uses, test the feasibility of the program plan, and provide a framework for design. The analysis is developed through a diagnostic process that cross-references the program plan with information generated in the site inventories. Optimal areas for programmatic elements are identified along with locations that may be too costly—environmentally, culturally, or economically—to develop.

A common site analysis method for synthesizing inventory data is to develop a series of informational maps—soil conditions, habitat type, zoning restrictions, or groundwater levels—that can be superimposed on one another and holistically examined. The overlays help to reveal relationships and patterns among site conditions. The maps can be generated on a transparent media and manually compiled; a more advanced and efficient method, such as GIS—a geographic information system—is also often used.

■ RESOURCES

Esri

http://www.esri.com/

GIS software, training, and support.

NRCS: maps, imagery, data, and analysis

http://www.nrcs.usda.gov/technical/maps.html

Library of land use, soil, wetlands, and other environmental resources

U.S. Census Bureau: Topologically Integrated Geographic Encoding and Referencing system

http://www.census.gov/geo/www/tiger/

Database containing digital features such as roads, railroads, rivers, lakes, and census information.

U.S. Geological Survey: maps and GIS data

http://water.usgs.gov/maps.html

Library of digital water information resources.

Establishing the Project Direction

Sustainable outcomes are more likely to come to fruition if they are considered from the outset of the project and pursued by the entire project team. Lack of alignment around a common purpose can create competing goals and priorities that impede the overall success of a project (7group and Reed 2009). Prior to starting the design process, a project team should work together to define the goals, design guidelines, and performance targets that will guide a project. Providing clear direction will keep the project team focused and encourage collaboration, allowing design solutions to evolve more quickly.

Project Goals

Goals are statements of intent that communicate what the project should realistically achieve. They convey the significance of a project, are action oriented, and are measurable. Projects often have multiple goals that define various components of the site. Goal-setting is a useful tool for establishing project direction and building consensus among team members.

Examples of project goals:

- Reduce crime and increase tourism by transforming the dilapidated site into a vibrant and sustainable community park.
- Design a safe and challenging outdoor environment for children that will encourage versatile play, creativity, and exploration of the natural environment.
- Create a backyard setting that provides habitat for songbirds and a place for quiet reflection.
- Transform the asphalt roof into a vegetated oasis that reduces stormwater runoff and provides building occupants with dynamic views that change seasonally.

Sustainable Design Guidelines and Performance Targets

Sustainable design guidelines are concepts that direct site development and outline the project approach. They provide a framework for the design to evolve within and are meant to guide rather than prescribe design solutions, leaving room for a wide range of creative options. Design guidelines reflect the unique characteristics and opportunities of the project program, site, and surrounding region. Many of the example guidelines outlined below are common criteria for green rating certification programs such as LEED, SITES, or BREEAM.

ENERGY	Minimize energy consumption and the generation of greenhouse gases.
PRODUCTS AND MATERIAL	Select materials and products whose extraction, production, transportation, use, and disposal minimize negative environmental and human health impacts.
	Follow the sustainable materials management hierarchy of (1) reduce material use, (2) reclaim and reuse materials, and (3) select materials that are made from recycled content and are recyclable. Limit waste disposal to toxic or dangerous materials.
	Design project elements so they can be deconstructed and reused in future projects.
	Design project elements in a manner that minimizes waste.
SOIL AND VEGETATION	Minimize the disturbance and removal of healthy soil and vegetation.
	Avoid development practices that increase the severity of natural disasters.
	Restore and reuse damaged soils.
	Protect and restore terrestrial and aquatic native plant communities and other site conditions that support wildlife habitat.
	Remove and avoid the use of exotic vegetation that is invasive to the region.
	Select vegetation that is well suited to the conditions of the site and can thrive with minimal and sustainable maintenance practices.
	Select vegetation that is resilient and can withstand the natural and human disturbances the site will receive.

WATER	Protect and restore natural water resources such as wetlands, streams, and rivers.
	Maintain or restore site appropriate hydrologic processes such as interception, infiltration, and evaporation.
	Capture, cleanse, and reuse stormwater and wastewater resources on-site.
	Minimize the use of potable water and other off-site water resources.
	Protect and restore native vegetative buffer zones along riparian, wetland, shoreline, and other water bodies.
	Protect and restore floodplain functions such as water storage, groundwater recharge, pollutant filtration, and wildlife habitat.
HUMAN HEALTH AND WELL-BEING	Create a unique site that reflects the local culture, materials, and vegetation.
	Provide opportunities for people to visually and physically connect with nature.
	Create landscapes that are inspiring and encourage a sense of stewardship.
	Design sites to promote physical, mental, and social health.
	Protect and maintain unique or historic site attributes.
EDUCATION AND CONTINUED LEARNING	Design the site to ease and encourage postoccupancy monitoring of sustainable design practices and visitor experiences.
	Make visible the rhythms and cycles of nature and the technologies and infrastructure that support the site's function.

Performance Targets

Performance targets should be established for each guideline to clarify desired outcomes and measures of project success. The targets are specific performance goals related to site sustainability that challenge the team to go beyond standard design criteria and reach a higher level of site performance. Once the performance targets are established, oversight of specific targets can be assigned to team members, who can then track progress and champion attainment. The targets serve as a common starting point for the design team; they may need to be adjusted as the project progresses and opportunities or constraints reveal themselves.

Examples of performance targets:

- Reuse or recycle 100 percent of the existing materials and vegetation found on-site.
- Reduce potable water use by 75 percent of the established baseline.
- Locate 100 percent of the soil displacement and disturbance on areas of the site degraded by previous development.
- Reuse 50 percent of the building's wastewater on-site.
- Create views of green space from 100 percent of the building windows.
- Reduce stormwater runoff by 90 percent.

Green rating certification programs such as LEED or SITES have established targets and specific documentation requirements for demonstrating criteria have been met. Whether or not a project is pursuing certification, a review of these targets can be helpful in establishing realistic and challenging goals.

Project teams hoping to achieve certification should familiarize themselves with the necessary parameters from the outset and clarify the roles and responsibilities of team members for documentation and certification. Deciding to pursue certification later in the process generally leads to increasing time and costs. Design teams should avoid selecting performance targets based solely on their point value and the short-term goal of certification.

Program Plan

A design solution that does not address the needs of the client is as ill-suited as one that does not properly fit the conditions of the site (Booth 1990). Clear communication and a thorough understanding of the needs and concerns of the client are key to a successful project. The program plan is a written description of the characteristics and requirements the design solution must satisfy. The plan clearly articulates the expectations of the client and guides the design team. Sustainability should be recognized as a necessary and integral component of the program plan; if a program does not directly address the desire to meet sustainable outcomes, it is unlikely to fulfill the requirements (Williams 2007).

The plan should be revisited throughout design and construction to ensure the project is progressing as envisioned. It is often necessary to revise the program to accommodate new challenges or criteria revealed by the design process. Any revisions should be discussed and agreed upon by the design team and the client, who is essential to the overall buy-in and long-term success of the project.

The program plan is developed through a multistep process described in more detail in the paragraphs that follow. The basic steps include:

1. Client interview
2. Workshop preparation
3. Program refinement workshop
4. Ongoing evaluations of design proposals to ensure they are in sync with the program plan

CLIENT INTERVIEW

The client interview is the first step in developing the program plan and includes feedback from the client, site users, and additional project stakeholders.

The following items should be discussed with the client and clearly documented:

- Project purpose

  Identify why the project is being built. What is it the clients would like to accomplish?

- Key decision makers

  Identify the individuals who will make the final decisions. Discuss how they will be involved in the design process.

- Site users

  Identify the site users, their age range, and any special requirements.

- Design elements and activities

  Make a list of the elements and activities site users need or desire. Outline the minimum and maximum requirements of each element and prioritize the list. Stay focused on the desired function of the site and do not get sidetracked with design options.

- Health benefits

  Identify the health benefits—physical, mental, and social—the design solution should offer.

- Environmental concerns

  Identify any regional or site-specific environmental concerns, such as air pollution or water shortages, that need to be addressed by the design solution.

- Educational opportunities

  Determine whether the client is interested in directed or experiential learning. Identify the audience.

- Aesthetic preferences

  Discuss the client's design style and aesthetic preferences. To help the client communicate the client's preferences, it can be helpful to provide visual examples. The intent is not to find a design solution, but to gain an understanding of the client's definition of beauty.

- Maintenance

  Outline the maintenance expectations. How much time and money would the client like to spend maintaining the site? Is the client interested in the physical activity benefits of doing the maintenance? Identify who will be responsible for the site maintenance. Are there any maintenance activities the client would like to avoid, such as pesticide use or mowing?

- Budget

  Identify the overall budget for the project, separating the initial investment, future phases, operations, and ongoing maintenance. The budget should be realistic but should not limit creativity. Ongoing dialogue with the client about the design solutions and their associated costs and benefits will be necessary throughout the design process.

WORKSHOP PREPARATION

Information gathered during the client interview is checked by the client for accuracy and shared with the design team. Collaboratively, the team reviews and discusses the interview and initial site inventory to prepare for the upcoming program refinement workshop. During this meeting, the team also identifies additional tasks or research necessary to support the workshop and assigns items to appropriate team members. Because no new information is being generated, it is not necessary for the client to be part of this discussion.

PROGRAM REFINEMENT WORKSHOP

During program refinement, the project team works with the client to define the project direction and potential in greater detail. The opportunities and constraints of the site are discussed, as are project goals and performance targets. Development of this final portion of the program plan typically takes place in a workshop or charrette setting and includes the client, all members of the integrated design team, and any additional site users or stakeholders.

The following items should be thoroughly discussed and documented:

- Site issues

  Discuss the findings of the initial site inventory and analysis. Identify any existing site issues that require careful evaluation and assign appropriate team members to gathering the information. If needed, discuss portions of the program that are not a good match or suitable for the site without major site changes, resources, or maintenance.

- Project goals

  Establish project goals.

- Sustainable design guidelines and performance targets

  List the guidelines and corresponding performance targets the team is striving to achieve. Discuss the relationship between the guidelines and the programmed design elements and activities. Identify the items that influence each other and create mutual benefits. Begin to consider which guidelines will be the easiest to achieve and which will require greater effort. Discuss which team members will champion the monitoring and successful attainment of the sustainable design guidelines. Identify the research and analysis needed to achieve the performance targets and begin conceptual design. Assign tasks to appropriate team members.

- Client interview

  Review the information gathered in the client interview, and determine whether any revisions need to be made due to recent discussions.

The finalized program plan provides clear direction and vision to keep the project focused as it evolves. Because the entire team has discussed key issues and developed clear direction, the project can unfold more rapidly, saving both time and money. With the program plan in place, the project team is now ready to begin exploring design solutions.

Developing Sustainable Design Solutions

Site design is the interface between natural and built systems. The blending of these systems requires an integrated design approach that is both creative and analytical. To achieve sustainable outcomes, the ecological, social, and structural components of the site must all be considered, as well as the relationship and influence between the components and the surrounding area. Design solutions are developed from an iterative process that cycles through phases of information gathering, analysis, and composition. The process continually builds upon itself, and as options are explored, progresses from the general to the more specific.

Outlined below are a series of distinct design phases and tasks that provide project teams with an organizational framework for sustainable site development. The design process is not linear and may need to vary depending on the unique circumstances of the project and desires of the client.

> Sustainable sites are not represented by any one design style or aesthetic; however, beauty is a very important aspect of sustainable design. Beauty draws people to the site and can provoke feelings of admiration and respect, which are vital to the development of an environmental ethic (Meyer 2008). Sites are beautiful when they function both ecologically and socially and inspire people to spend time outdoors and connect with nature.

Conceptual Diagrams

With the project intent and expectations clearly defined in the program plan, the team is ready to begin studying design possibilities in graphic format. Conceptual diagrams explore the approximate location and relative size of programmatic elements in relation to the site.

In this creative and open process, multiple scenarios are explored and analyzed to help the project team study the site from various perspectives and identify those that have the greatest potential. Freehand drawings are developed somewhat quickly, typically using bubbles, symbols, and hatch patterns to communicate ideas. The entire site area should be considered and assigned a designated use, leaving no blank areas or holes in the diagram. At this phase in the design process, no specific shapes or forms are studied. The diagram is drawn on an overlay of the site analysis and base map drawings in an effort to unify the program plan with the existing site conditions.

Conceptual Diagram: Design Guidance for Sustainable Outcomes

REVIEW MATERIALS AND SHARE PROGRESS

- Review the program plan with the design team. Renew commitment to the project goals, sustainable guidelines, and performance targets.
- Review the site inventory and analysis in detail and discuss the opportunities and constraints of the site. All members of the design team should have a thorough understanding of the existing site conditions and, if at all possible, have visited the site prior to starting design.
- Share the findings of the research and analysis done to support the conceptual diagram phase. Detailed research and analysis of the site and proposed systems will aid in the optimization of the design solutions.

SITE LAYOUT

- Study the spatial relationship between programmatic elements. Determine what elements need to be close together and which should be separated.
- Consider the space requirements for the programmatic elements and the required maintenance activities.
- Brainstorm strategies for achieving performance targets. This discussion is not intended to be limiting or to forge a commitment to any one strategy, but to serve as a starting point for design.
- Look for opportunities to frame desirable views and screen on-site and surrounding features that are unsightly or noisy.

continues

Conceptual Diagram: Design Guidance for Sustainable Outcomes

SITE LAYOUT *(CONTINUED)*

- Create a sun-path diagram for the site and study the shadowing from trees, topography, and structures in the summer and winter months. Orient buildings to reduce energy use and outdoor gathering space to take advantage of comfortable microclimates.
- Identify existing structures and hardscape areas that are to remain and those that will be disassembled for reuse or recycling.
- Look for opportunities to link the site to pedestrian and bicycle networks.
- Study the existing topography and locate programmatic elements to minimize the disturbance and removal of healthy soil and vegetation.
- Identify existing cultural and natural resources that need to be protected or restored in order to achieve the project's goals and performance targets.
- Consider reuse and restoration options for portions of the site that are ecologically degraded, such as compacted soils or areas of invasive vegetation. Degraded areas should be considered first for design elements that will require significant soil and vegetation disturbance.
- Provide adequate space to protect and restore native vegetative buffer zones along riparian, wetland, shoreline, and other water bodies.
- Study surrounding land uses and resources. Explore opportunities to link to off-site habitats and extend corridors through the site.
- Identify existing and potential stormwater runoff sources and begin to explore options to capture, cleanse, and reuse the water on-site.

ASSIGNMENTS AND NEXT STEPS

- Identify the items that need to be researched and analyzed in greater detail before beginning the schematic design phase. Consider the information required to successfully attain the project goals and performance targets. Assign tasks to appropriate team members.
- Review and refine project schedule.

Schematic Design

Schematic designs are a series of drawings that build upon the conceptual diagram and add form and detail to the design solution. In this phase, the project team collaboratively addresses problematic site issues and explores options for optimizing and elegantly integrating design solutions into the site. The majority of the design work is accomplished during this phase.

A number of design alternatives are considered and the benefits and tradeoffs of each explored. Multiple design studies that utilize the full expertise of the multidisciplinary team foster a greater understanding of the project's constraints and opportunities. Special attention should be given to protecting or restoring the ecological and cultural integrity of the landscape. The design team should keep an open mind and avoid settling on a definite direction too early. Taking time for reflection, reinvestigation, and research will allow new ideas to develop and be successfully implemented (7group and Reed 2009). Hybrid designs will often emerge that bring together the best ideas from each plan.

Any deficient or unrealistic components of the program plan are often revealed at this stage and will need to be reconciled with help from the client and other team members. Design solutions should be evaluated against the program plan and progress toward achieving the project goals and performance targets tracked. Sustainable design strategies not included in schematic drawings can often be lost in future phases (Kwok and Grondzik 2007). Assumptions and standard design approaches that work against sustainable solutions need to be thoughtfully challenged (Mendler, Odell, and Lazarus 2006). Proceeding without questioning assumptions leads to lost opportunities.

During this phase, preliminary cost estimates should be established. Design changes made in schematic are easier and less expensive to make than they are in future project phases. It is often assumed that the additional research and analysis required in the integrated design process will slow progress and increase project costs; however, this is not necessarily the case. The emphasis on research and analysis in schematic allows the design development and construction documentation phases to be significantly reduced, as these phases become more about fine-tuning the design and documentation for construction and are not encumbered by continual redesign efforts (7group and Reed 2009).

Schematic Design: Design Guidance for Sustainable Outcomes

REVIEW MATERIALS AND SHARE PROGRESS

- Review the program plan, project goals, sustainable guidelines, and performance targets.
- Share findings of the research and analysis gathered to support the schematic design phase.

SOIL AND VEGETATION

- Explore options for the landscape to provide physical, mental, and social health benefits to the site users. Provide experiences that connect people to nature and build an environmental ethic.
- Minimize disturbance of healthy soils and vegetation. Artfully incorporate existing vegetation and topography into the design solution.

continues

Schematic Design: Design Guidance for Sustainable Outcomes

SOIL AND VEGETATION *(CONTINUED)*

- Identify the existing vegetation that will be removed. Explore options to reuse (transplant) or recycle (i.e., mulch or vegetative bundles for erosion control) on-site.

- Consider reuse and restoration options for the portions of the site that are ecologically degraded, such as compacted or eroded soils and areas dominated by invasive species. Degraded areas should be considered first for design elements that will require significant soil and vegetation disturbance.

- Strategically locate vegetation and vegetated structures, such as trees and shade trellises, to create comfortable microclimates and reduce the energy consumption of surrounding buildings.

- In fire-prone areas, lay out the site, design structures, and select vegetation to reduce the risk of damage or loss due to wildfire. Research Firewise construction guidelines.

- Consider options to grow food for the site users and others.

WATER

- Explore opportunities to capture, reuse, and recycle all available water resources on-site, including rainwater, stormwater, greywater, air-conditioner condensate, and wastewater. Run preliminary calculations to determine the amount of water available and discuss creative solutions that allow the water to be used safely on-site. Discuss local codes and process for attaining permits.

- Work with the building architect to include opportunities for greywater, air-conditioner condensate, and wastewater collection in the building design.

- Minimize impervious surfaces. Consider options for covering or shading surfaces with vegetation such as green roofs, trellises, green walls, and arbors.

MATERIALS

- Follow the sustainable materials management hierarchy of (1) reduce material use, (2) reclaim and reuse materials, and (3) select materials that are made from recycled content and that are recyclable.

- Consider the environmental and human health impacts of material extraction, production, transportation, and disposal. Consult life-cycle assessment tools such as the Athena Environmental Impact Estimator or the National Institute of Standards and Technology Building for Environmental and Economic Sustainability (BEES).

Schematic Design: Design Guidance for Sustainable Outcomes

MATERIALS *(CONTINUED)*

- Use minimally processed materials such as uncut stone, earth materials, and wood.
- Select products and materials that are durable, have a long life, and can be easily reused in future projects.
- Research options for products and materials made from rapidly renewable resources. Consider the durability and estimated life span of the product.
- Determine which materials are recyclable within the project area. Identify recycling centers and their requirements for accepting materials.
- Research options for regionally extracted and manufactured materials.
- Work with the building architect to locate HVAC in an area that encourages energy efficiency and reduces noise impacts to the landscape. If HVAC cannot be moved, research options to insulate the sound and shade the unit.
- Run preliminary calculations to determine the site's energy requirements. Research options to reduce energy consumption and produce or purchase renewable energy.

ASSIGNMENTS AND NEXT STEPS

- Begin development of the site monitoring and maintenance plan.
- Identify the equipment or information needed to monitor the site's performance and track sustainable design features. Consider how monitoring mechanisms will be included in the design and how the information will influence the ongoing management of the site.
- Validate that the schematic design meets the performance targets before moving on to design development and engaging in more detailed design and optimization.
- Identify the items that need to be researched and analyzed in greater detail before beginning the design development phase. Consider the information required to successfully attain project goals and performance targets. Assign tasks to appropriate team members.
- Prepare preliminary cost estimates and analysis. Cost-saving opportunities typically decrease as a project progresses and design changes become more costly.
- Review and refine the project schedule.

Budget

Early efforts to investigate and manage the budget offer the best opportunities to optimize the monetary resources available to the project. Maximum cost savings are achieved when sustainable design strategies are incorporated from the outset of the project in collaboration with an integrated design team (Kubba 2010). The more detailed design solutions become without exploring costs and potential savings opportunities, the more costly it will be to make adjustments to them.

It is a common misconception that sustainable design automatically leads to increased costs. In many cases in which sustainability has been incorporated into the foundation of the project, innovative solutions can be found within the project budget (Mendler, Odell, and Lazarus 2006). This is often a surprise to those accustomed to the traditional design process and who expect sustainability to be the product of adding green technologies and materials to the design solution once it is established (7group and Reed 2009).

It is the responsibility of the design team to help the client understand the full environmental, human health, and social costs of a project, many of which go beyond initial design and construction expenses. Outlined below are items that will aid the design team in developing a more holistic cost perspective and achieving sustainable outcomes within the project budget.

- Set a project goal of obtaining a zero-cost increase over a standard budget for similar project types to encourage cost management in all project phases (Yudelson 2009).
- Evaluate the expense of project components from a whole-system perspective. Understand their relationship to one another and how they influence the performance or success of other components of the design. Avoid the temptation to reduce the quality of a component based solely on its line item cost. Organize the budget into performance "bundle" costs, grouping together project components that influence the performance of one another. Consider how reducing or removing a component may require increased expenses in other areas during construction or during the life of a project (7group and Reed 2009).
- Conduct life-cycle cost analysis (LCCA) to evaluate all relevant costs over the life of the project or an individual product. Calculations typically include: the initial investment; ongoing operation; utilities such as energy, water, or waste disposal; maintenance and monitoring requirements; capital replacement costs; and disposal costs minus the salvage value. LCCA is a useful tool for comparing competing project alternatives and maximizing new savings.
- Research existing rebates, or programs that can help offset costs or provide savings, such as solar panels or rainwater-harvesting barrels.
- Research grants, tax incentives, and other benefits associated with specific locations or project types, such as urban infill, greyfield, or brownfield sites.
- Consider environmental and human health cost and benefits.
- Explore opportunities to build the project in phases to accommodate immediate and future budgets. Design project components and systems to support the successful expansion and completion of the master plan.

Project spending should be prioritized to support design solutions that have multiple long-lasting benefits and create the greatest environmental and human health gains. Include the entire design team in budget discussions and utilize the group intellect to determine optimal ways of cutting costs and maintaining project performance. To protect the initial investment and ensure successful project completion, funding should be allocated to allow the project team to be part of the construction process and develop a site maintenance and monitoring plan.

Cost estimating is an important component of each phase of the design process. Exploring costs while working through the design will help the team find affordable and sustainable solutions that are within the project budget and avoid shortsighted value engineering efforts.

Design Development

Design development is the last phase of the design process. Building on schematic drawings, it focuses on the detailed appearance, exact size, and optimal function of landscape elements and materials. Because the major design decisions have been made in the schematic phase, design development can be centered around the fine-tuning and optimization of design solutions to achieve multiple and long-lasting benefits. Front-loading design exploration and research in the conceptual and schematic phases allows the integrated design team to establish design solutions early in the process, leaving time for a higher level of analysis and detail to occur in design development (7group and Reed 2009).

Design Development: Design Guidance for Sustainable Outcomes

REVIEW MATERIALS AND SHARE PROGRESS

- Review the program plan, project goals, design guidelines, and performance targets. Assess the realistic potential and renew commitment.
- Validate that the schematic design meets the project program requirements, including the goals and performance targets, before engaging in more detailed design and systems optimization.
- Discuss the relationships among design components. Identify how the performance of a component is dependent on other portions of the design.
- Share findings of the research and analysis gathered to support the design development phase.
- Review the budget estimates. Include the entire design team in budget discussions and utilize the group intellect to determine optimal ways of cutting costs and maintaining project performance.
- Review the draft monitoring and maintenance plan. Determine if design changes are needed in order to create more successful monitoring and maintenance outcomes.

continues

Design Development: Design Guidance for Sustainable Outcomes

SOIL AND VEGETATION

- Minimize grading and balance cut and fill.
- Select vegetation that is well adapted to the site conditions.
- Select vegetation that can thrive without the continued use of potable water. Fully utilize on-site alternative water resources such as rainwater, air-conditioner condensate, stormwater, greywater, and wastewater.
- Use a diverse plant palette appropriate for site conditions. Avoid large expanses of monocultures. Give preference to plants native to the region.
- Reuse (transplant) or recycle (i.e., mulch or use vegetative bundles for erosion control) all existing vegetation on-site whose presence will not cause harm or risk to the site (such as diseased or invasive vegetation). Transplanting may be particularly important for native species that are not readily available and cannot be easily replaced.
- Avoid the use of plants invasive to the region.
- Use native plant communities as models for plant palettes.
- Select vegetation that provides a source of food for humans and/or wildlife.
- Research opportunities to recycle and reuse all organic matter generated during site maintenance.
- Document requirements and necessary steps for protecting or restoring existing natural resources such as soils, vegetation, and wildlife habitat.
- Select vegetation that is resilient and can withstand the natural (floods, fires, high winds, etc.) and human (pedestrian traffic) disturbances.
- Detail site features in a way that connects people to nature and provides physical, mental, or social health benefits.

Design Development: Design Guidance for Sustainable Outcomes

WATER

- Optimize stormwater, greywater, and wastewater management systems. Strive to reuse all water on-site.

MATERIALS

- Follow the sustainable materials management hierarchy of (1) reduce material use, (2) reclaim and reuse materials, and (3) select materials that are made from recycled content and are recyclable.
- Design landscape elements so that they can be deconstructed and reused in future projects.
- Design landscape elements to be consistent with the standard size of materials and minimize additional cuts and waste.
- Use sustainable certified products.
- Avoid the use of resources that are nonrenewable or regenerate slowly, such as sphagnum peat.
- Eliminate the use of wood from rare, threatened, or endangered trees.
- Select energy-efficient fixtures and equipment.
- Use nontoxic, organic, or natural materials and products. Avoid products that off-gas and release harmful levels of volatile organic compounds (VOCs) and other chemicals while on site or during manufacturing or disposal.
- Use lighting efficiently and accurately to increase safety and reduce light pollution.
- Select paving and roofing materials with a solar reflectance index of at least 29 to reduce urban heat island effects.

continues

Design Development: Design Guidance for Sustainable Outcomes

ASSIGNMENTS AND NEXT STEPS

- Validate that the design development package fulfills the requirements of the program plan.
- Identify the items that need to be researched and analyzed in greater detail before beginning the contract document phase. Consider the information required to successfully attain project goals and performance targets. Assign tasks to appropriate team members.
- Conduct budget estimates and analysis.
- Continue to develop the monitoring and maintenance plan.
- Review and refine the project schedule.

Construction Documents

The design team should strive to focus this phase of the project on documenting the design, not designing while documenting. If the integrated design process has been successful, the coordination and synthesis of the project components has already occurred in previous phases and is built into the design. As a result, fewer errors and omissions generally occur in the contract documents, and change orders can be significantly reduced (7group and Reed 2009). If design decisions are still being made while trying to develop contract documents, there is a great chance errors will be made and opportunities to optimize systems and control costs effectively will be lost.

The contractor is instrumental to the successful completion of a project and should understand and embrace their role in helping the project achieve sustainable outcomes. If the contractor was not part of the integrated design process, they will not be aware of the intent and commitments behind the documents and will be more likely to propose substitutions or changes that might alter the original purpose (Kwok and Grondzik 2007). In this situation, the project goals, performance targets, and reasoning behind the design choices should be discussed in detail with the contractor and subcontractors—not only the supervisors but also construction personnel who will actually be working on the job site—to ensure clear communication and follow-through.

Construction Documents: Guidance for Sustainable Outcomes

REVIEW MATERIALS AND SHARE PROGRESS

- Verify the achievement of program plan requirements, including the goals and performance targets, before beginning the development of construction documents.
- Discuss the relationships between design components. Identify how the performance of a component is dependent on other portions of the design.
- Share findings of the research and analysis gathered to support the construction documentation phase.
- Discuss the content that needs to be included in the construction documents and how to integrate and communicate details so that the project can be accurately priced and constructed.
- Review the budget estimates. Include the entire design team in budget discussions and utilize the group intellect to determine optimal ways of cutting costs and maintaining project performance.
- Review and finalize the monitoring and maintenance plan.

CERTIFICATION AND SCHEDULE

- If pursing certification, specify the requirements and responsibilities of the contractor in the certification process.
- Ensure the schedule allows adequate time for the deconstruction of structures and amenities and the removal of vegetation so that they are not destroyed and can be reused or recycled.

DRAWINGS AND SPECIFICATIONS

- Develop a site protection plan that minimizes clearing, grading, and other site disturbances. Communicate the limits of construction and all areas to be fenced and protected throughout construction. Dictate monetary consequences for damaging the site beyond the agreed-upon construction envelope.
- Provide a specific location for storing equipment, stockpiling materials, travel routes, and parking areas for construction equipment.
- Specify the requirements necessary for achieving the sustainable guidelines and performance targets, such as the appropriate disposal of invasive species, tree protection, soil restoration techniques, and construction waste recycling. Include the appropriate verification and testing methods to confirm components and systems are properly functioning.

continues

Construction Documents: Guidance for Sustainable Outcomes

DRAWINGS AND SPECIFICATIONS *(CONTINUED)*

- Develop a plan for the safe use and transportation of chemicals, fuels, and other hazardous materials. Minimize the storage of these materials on-site. When on-site storage is necessary, identify the appropriate holding areas and methods. Research local transportation and use regulations.

- Specify energy performance requirements for equipment, such as lighting, irrigation, and control systems.

- Prior to beginning construction or demolition work, require the contractor to submit a construction waste management plan. The plan should include the materials to be recycled, estimated quantities, cost comparison of recycling and disposal, transportation methods, and names of licensed recycling centers that will receive the materials (Mendler, Odell, and Lazarus 2006).

- Avoid the specification of automatic irrigation systems that do not take into consideration current site conditions and the needs of the vegetation.

MANUFACTURERS

- Request recommendations from the manufacturers for preferred maintenance methods that have the least environmental and human health impacts (Mendler, Odell, and Lazarus 2006).

- Select manufacturers who have take-back programs and reuse or recycle their packaging and product.

CONSTRUCTION EQUIPMENT

- Specify construction equipment that reduces fuel consumption and the release of greenhouse gases.

- Require the contractor to implement an idle-reduction policy that reduces emissions from construction equipment by limiting unnecessary idling to no more than five minutes in any sixty-minute period (Sustainable Sites Initiative 2009).

- Specify construction equipment and practices that reduce damage to the site. Avoid the use of oversize maintenance equipment that causes soil compaction and vegetation loss. Use the smallest and lightest tools that can accomplish the job (Thompson and Sorvig 2000).

Construction Documents: Guidance for Sustainable Outcomes

SOIL, VEGETATION, AND MATERIALS

- Schedule soil disturbance and the removal of vegetation to the smallest practical space to minimize the area exposed at any one time during construction. Soil should not be left unnecessarily bare for an extended period of time.
- Provide the necessary erosion-control measures to protect the soil throughout the construction process.
- Restore soils disturbed during construction in all areas that will be revegetated.
- Remove and dispose of existing invasive plants in a manner that does not encourage their spread.
- Specify the purchase of vegetation and materials from the following distances:

  Soil and aggregate materials that have been extracted, harvested, or recovered and manufactured within 50 miles of the project site.

  All growing facilities for vegetation located within 250 miles of the project site.

  All other materials should be extracted, harvested, or recovered and manufactured within 500 miles of the site.

Construction Observation

The integrated design process does not end with the development of construction documents. Coordination and collaboration continue through the construction phase to ensure the project meets its goals and performance targets. Regular on-site observations are required to monitor progress, implement quality control measures, and address unforeseen scenarios.

Strategies for encouraging collaboration and achieving project goals and performance targets throughout construction include the following:

- Develop a relationship with the contractor that focuses on collaboration toward mutual goals, as opposed to a feeling of oversight or policing of their work.
- Reiterate project goals and performance targets with the contractor and subcontractors throughout the construction process.
- Require contractors to track and report progress toward meeting goals and performance targets.
- Use site visits as opportunities to communicate the basis for design decisions and their envisioned end results.
- Test systems and equipment under multiple scenarios to ensure they are assembled, installed, and operating correctly; this is particularly important for new or unique systems that the contractor may not be as familiar with or that require multiple trades to construct.
- Attend regular meetings with the contractor to continue an open dialogue and problem-solving relationship.

Site Monitoring and Maintenance: Developing an "Owners Manual"

Sites that are not properly maintained result in an unsustainable cycle of "remove, replace, and rebuild" that increases environmental and economic costs. For a project to avoid these costs and fulfill its goals over the long term, the site's maintenance must be considered from the outset. Integrating a land-care professional into the design process allows maintenance requirements to be discussed throughout the development of the project and helps to ensure that the landscape can be sustainably cared for within the client's available resources.

Because landscapes are comprised of living systems that evolve and change over time, informed and intentional stewardship is required to maintain the ecological and cultural integrity of the site. A site monitoring and maintenance plan is a necessary component of a successful project and should be used to convey the activities and schedule required to support the project goals and performance targets over the life of the project. The intention of the plan is not to maintain a static landscape, but to guide the evolution and adaptation of the site in a way that continually improves ecological function and the visitor's experience. Postoccupancy evaluations and monitoring of the biophysical conditions and sustainable design practices help to ensure that the stewardship of the site will be truly effective.

The monitoring section of the plan outlines strategies and procedures for tracking success, identifies red flags, and informs the maintenance personnel about how to interpret and use the information to improve the function of the site. The plans are a valuable asset that can be passed on to future owners, to help ensure the continued success of a project, and should include the information and background materials necessary to support such a transition.

A sample list of items typically covered in a monitoring and maintenance plan is included below.** Monitoring and maintenance plans are unique to the design solution and may vary between projects.

PROJECT BACKGROUND

- Purpose
- Short- and long-term project goals
- Performance targets
- As-built construction documents

PLANT STEWARDSHIP

- Monitoring guidance and instruction on how the collected data informs the adaptive management of the site
- Routine maintenance
- Methods for diseased plant disposal
- Replacement criteria
- Invasive species management

SOIL STEWARDSHIP

- Monitoring and soil-testing guidance and instruction on how the collected data informs the adaptive management of the site
- Routine maintenance
- Erosion and compaction prevention and management

**Sustainable Sites Initiative 2009.

WATER STEWARDSHIP

- Monitoring guidance and instruction on how the collected data informs the adaptive management of the site
- Irrigation requirements
- Natural and constructed water features maintenance
- Stormwater, greywater, and wastewater management features and system requirements

MATERIALS STEWARDSHIP (INCLUDES ALL HARDSCAPE AND STRUCTURES)

- Monitoring guidance and instruction on how the collected data informs the adaptive management of the site
- Care and replacement
- Site safety
- Disposal of harmful materials
- Expected energy use

EQUIPMENT SELECTION AND STEWARDSHIP

- Low-emission maintenance equipment
- Care and replacement
- Site safety
- Anticipated energy consumption and monitoring requirements

CLIMATIC CONDITIONS

- Snow or ice removal
- Flooding cleanup and repair

Site Monitoring and Maintenance: Guidance for Sustainable Outcomes

MAINTENANCE PLAN

■ Prescribe monitoring practices to ensure the site is functioning as envisioned. Provide guidance on how to interpret the information and use it to inform the adaptive management of the site. If problems are identified, provide instructions for how to resolve them or who to contact.

■ Document the appropriate maintenance methods and schedules required to support the project goals and performance targets. Communicate expectations for how the landscape will mature and evolve over time.

continues

Site Monitoring and Maintenance: Guidance for Sustainable Outcomes

MAINTENANCE PLAN *(CONTINUED)*

- Recommend equipment that reduces fuel consumption and the release of greenhouse gases. Avoid oversize maintenance equipment that causes soil compaction and vegetation loss. Recommend the smallest and lightest tools that can accomplish the job. Include equipment manuals and maintenance requirements.
- Research the maintenance requirements of materials. Specify nontoxic or the least toxic options, and provide a maintenance schedule.
- Incorporate integrated pest management practices.
- Include a list and photos of the common invasive species in the region that may volunteer on the site. Document the appropriate removal methods or resources for more detailed information.
- Encourage the client to reuse vegetation trimmings on-site as compost or mulch.
- Require that the soil be tested for deficiencies prior to amending. Discourage the use of quick-release fertilizers. Encourage the use of compost in lieu of fertilizers.
- Track water use and schedule regular inspections of irrigation equipment and other water features for breaks, leaks, and general malfunctions.
- Educate the client about the amount of water required to support the vegetation on-site and why it is important to irrigate only when needed. Encourage the client to water manually or observe the irrigation system when it is running to reduce water waste.
- Specify snow and ice removal methods that do not harm vegetation and are not harmful to the surrounding landscape. Identify the appropriate locations for snow piles.
- Provide instruction on how to monitor the site's energy use and look for opportunities to reduce consumption.
- Encourage the client to purchase energy generated from renewable sources. Research local energy options and provide the client with the necessary information to make sustainable energy choices.
- Review the plan with the client and other individuals responsible for the care of the landscape. Address any questions or gaps in the plan. Emphasize the importance of their work in helping the project continue to meet its goals over the long term.
- Schedule monitoring and maintenance plan reviews and adjustments on an annual basis, or as needed.

REFERENCES

7group and B. Reed. 2009. *The integrative design guide to green building.* Hoboken, NJ: John Wiley & Sons.

American Society of Landscape Architects. 2009. Positive economic news continues for landscape architecture firms [press release].

Booth, N. 1990. *Basic elements of landscape architecture design.* Prospect Heights, IL: Waveland.

Keeler, M., and B. Burke. 2009. *Fundamentals of integrated design for sustainable building.* Hoboken, NJ: John Wiley & Sons.

Kubba, S. 2010. *Green construction project management and cost oversight.* Burlington, MA: Elsevier Architectural Press.

Kwok, A. W., and W. Grondzik. 2007. *The green studio handbook: Environmental strategies for schematic design.* New York: Architectural Press.

LaGro, J. A. 2008. *Site analysis: A contextual approach to sustainable land planning and site design.* Hoboken, NJ: John Wiley & Sons.

Mendler, S., W. Odell, and M. A. Lazarus. 2006. *The HOK guidebook to sustainable design.* Hoboken, NJ: John Wiley & Sons.

Meyer, E. K. 2008. Sustaining beauty: The performance of appearance. *Journal of Landscape Architecture* (Spring).

Sustainable Sites Initiative. 2009. *Sustainable sites initiative: Guidelines and performance benchmarks 2009.*

Thompson, J. W., and K. Sorvig. 2000. *Sustainable landscape construction: A guide to green building outdoors.* Washington, DC: Island Press.

Williams, D. E. 2007. *Sustainable design: Ecology, architecture, and planning.* Hoboken, NJ: John Wiley & Sons.

Yudelson, J. 2009. *Green building through integrated design.* New York: McGraw-Hill.

© MITHUN | JUAN HERNANDEZ

■ FIGURE 3.1

CHAPTER 3

Human Health and Well-Being

DISCUSSIONS OF HEALTH often focus on illness; however, human health is more than the absence of disease or infirmity: it is a state of complete physical, mental, and social well-being (World Health Organization 2010). Sustainable sites help people live healthier, fuller lives by supporting all aspects of human health and by integrating opportunities for safe and convenient physical activity, social interaction, and mental restoration into our daily routine.

Physical Health

Regular physical activity is vital to maintaining a healthy weight and reducing the instance of disease. Sedentary habits and obesity increase the risk of heart disease, stroke, high blood pressure, osteoarthritis, gall bladder disease, diabetes, and some cancers (Centers for Disease Control 2008). The lack of physical activity among children and adults has become so critical it is considered to be a major health risk in the United States and many other developed nations (see Figure 3.2). Obesity in children is particularly concerning due to linkages with the early onset of chronic illnesses. Doctors predict that because of obesity, the current generation of children in the United States will, on average, live less healthy and shorter lives than their parents (Olshansky et al. 2005).

Sites can improve human health by providing opportunities for safe and convenient physical activity. Early in the design process, project teams should identify the interests, abilities, and preferred physical activities of user groups, as well as the opportunities for connecting to surrounding sidewalks, trails, bicycle networks, and other sites.

People are more likely to live an active lifestyle when it is a part of their everyday routine. Projects that encourage physical activity not only help site users reduce the instances of disease but also boost energy levels, improve mental health, help prevent depression, and maintain self-esteem.

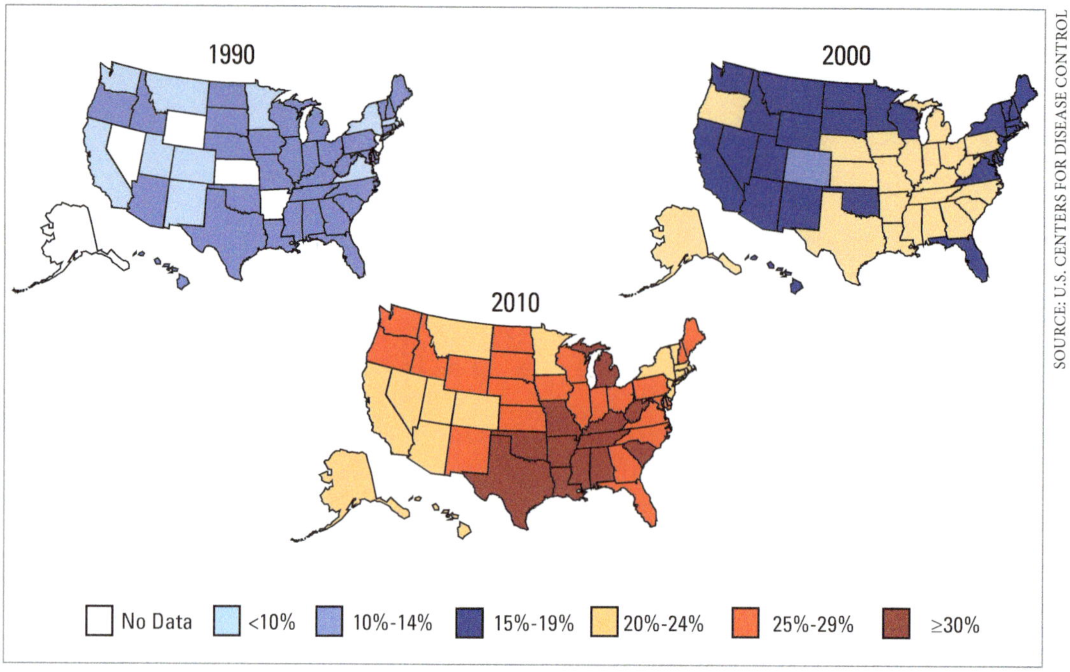

■ **FIGURE 3.2** Obesity trends among adults in the United States have dramatically increased over the past 20 years. The lack of physical activity among children and adults has become so critical it is considered to be a major health risk in the United States and many other developed nations.

Strategies for increasing physical activity through site design include the following:

- Make outdoor physical activity convenient and inviting.
    - Link the site to community and regional sidewalk and trail systems.
    - Provide attractive landscape elements that encourage walking. For example, a housing development may locate mailboxes in a central area that is easily accessible by foot, or an office complex may locate a coffee vendor in a centrally located garden space.
    - Limit parking and provide amenities, such as covered bike racks, water fountains, showers, and changing rooms, to encourage walking or biking to the site.
    - Design spaces to be multifunctional and capable of supporting organized or impromptu physical activity.
    - Encourage gardening as a physical activity. Basic gardening activities can burn an average of 300 calories per hour for a 150-pound (68 kg) person (Calorie Count 2010).
- Provide for user safety and comfort.
    - Protect site users from adverse climatic conditions. Provide options for physical activity in both sun and shade.
    - Implement Crime Prevention through Environmental Design (CPTED) or similar safety design strategies.
    - Design walkways and trails to be visible and easily accessible from nearby buildings, streets, and other activity areas.
    - Provide a variety of entrances and exits to allow site users to control their experience and safely chose alternative routes.
    - Avoid routing trails near vehicular lanes or roadways.
    - Construct wide and unobstructed pathways. Provide seating opportunities at regular intervals.
    - Create an environment that is easy and intuitive for site users to navigate.
    - Provide open sight lines and view corridors.

- Redesign underutilized outdoor spaces.
    - Convert vacant lots into community gardens or parks.
    - Replace underutilized parking or roof space with gardens. Provide opportunities for site users to manually maintain and cultivate the landscape, such as flower or vegetable gardens, compost piles, or manual mowers.

■ RESOURCES

Active Living by Design, Inc.

www.ActiveLivingResearch.org

Centers for Disease Control and Prevention: Physical Activity Tool Kits

http://www.cdc.gov/nccdphp/dnpao/hwi/toolkits/physicalactivity.htm

■ CASE STUDY

ACKERMANNBOGEN NEIGHBORHOOD PARKLAND

PROJECT TYPE: Park

LOCATION: Munich, Germany

SIZE: 6 acres (2.4 hectares)

COMPLETION DATE: 2008

CLIENT: City of Munich, Department of Public Construction

HIGHLIGHTED SUSTAINABLE PRACTICES:

Local district heating from renewable resources

Reuse of on-site waste materials

Zero stormwater runoff

Provides opportunities for physical activity and socialization

Provides habitat for blue butterflies, partridges, and other endangered species

THE SITE: Former army barracks, recreational fields, and gravel overflow parking lot

M. REUSS, BAVARIAN CENTER FOR APPLIED ENERGY RESEARCH / ZAE BAYERN

■ **FIGURE 3.3**
Construction of the 52-foot-high (16 m), 85-foot-wide (26 m) tank, which stores 4.7 acre feet (6,000 m^3) of hot water heated by solar panels located on nearby apartment rooftops. Soil insulates the storage tank, which provides 50 percent of the heat and hot water demand for 320 homes. The solar hot water system is integrated into the design of the park and doubles as a popular sledding hill.

Design Overview

The public parkland project is part of a greenbelt that surrounds a 97.6-acre (39.5-hectare) mixed-use redevelopment. The park includes open space, recreational areas, trails, and a sledding hill that insulates an innovative hot water storage tank. The solar heating system is a pilot project that aims to produce environmentally friendly energy for part of the residential area. Heat is collected during the summer through large solar panels, which cover the roofs of adjacent apartment blocks, and is transferred via insulated pipes to the water storage tank. The innovative system provides hot water for 320 apartments throughout the year and also 50 percent of their heating needs during the winter months (see Figure 3.3).

continues

ACKERMANNBOGEN NEIGHBORHOOD PARKLAND *(CONTINUED)*

The design was inspired by the adjacent landscape of the Munich Olympic Village, which is one of the city's major landmarks. Shops, schools, and other necessary infrastructure are within walking or cycling distance of the development, and the park's attractive foot and cycle paths encourage physical activity and the use of nonmotorized transport.

GABRIELLA ZAHARIAS

■ **FIGURE 3.4**
Community playground with sledding hill in the background.

Approximately 26,159 cubic yards (20,000 m^3) of gravel was removed during the construction of the neighboring buildings and was reused in the park to sculpt the landscape. The site was designed to be a carbon sink and reduce the urban heat island effect. Native trees and green areas increase the total vegetative biomass of the city and are supported by sustainable drainage structures that allow rainwater to seep directly into the soil. The north portion of the park is planted with ecologically appropriate vegetation typical of the heath landscape around Munich. The dry meadow acts as a stepping-stone corridor for the habitat of blue butterflies, partridges, and other endangered species.

ACKERMANNBOGEN NEIGHBORHOOD PARKLAND *(CONTINUED)*

GABRIELLA ZAHARIAS

PROJECT TEAM

- **LANDSCAPE ARCHITECTS**

  zaharias landschaftsarchitekten: Gabriella Zaharias

  http://www.zaharias.net

  Matthias Thoma

  http://www.thoma.la

- **LOCAL CONSTRUCTION SUPERVISION**

  Walter Zimmermann, Landscape Architect

■ **FIGURE 3.5**
Children enjoy the challenge of climbing to the top of a long slide embedded into the site topography.

Mental Restoration

On a daily basis, people face numerous demands that can result in mental fatigue and stress. Under such conditions, we can suffer from irritability, physical tiredness, an inability to concentrate, and weakened immune systems. In the United States, workplace stress alone has been found to cost more than $300 billion each year in healthcare, diminished productivity, and employee turnover (American Institute of Stress 2010).

Connecting with the natural environment, whether by physically going out into the landscape or by looking at a garden view through a window, can provide a variety of mental and physical health benefits (see Figure 3.6). Researchers at the University of Illinois Landscape and Human Health Laboratory have found direct access to green space to be associated with lower levels of irritability and aggression and an improved ability to concentrate (Kuo and Sullivan 2001). Similar studies conducted by Rachel Kaplan (1993) and others found office workers with views of trees and vegetation to be more productive, to have less absenteeism, and to be generally more satisfied. And in studies that are changing the design of healthcare facilities, Roger Ulrich (1984) found patients with views of vegetation and natural settings recover from surgery more quickly, require less pain medication, and have fewer complications.

The link between nature and improved mental function and coping abilities has been attributed to the opportunities natural settings provide to relax and renew our minds and bodies. Project teams wanting to provide opportunities for mental restoration should consider the landscape views from within the surrounding buildings as well as spaces within the landscape. Successfully designed sites make visual and physical access to nature an integrated and unavoidable part of the design. In doing so, they improve the mental health and overall capacity of site users to manage major life issues (Kuo and Sullivan 2001).

Strategies for providing restorative settings include the following:

- Frame and direct views to wilderness or garden areas.
- Screen views of electrical transmission towers, HVAC equipment, power lines, prominent concrete or asphalt surfaces, and other artificial elements.

FIGURE 3.6 All residents of the Arkadien Asperg multifamily development have views and access to green space. The rainwater-fed stream is an enjoyable highlight of the extensive stormwater management system that meanders through the site.

DIETER GRAU / © ATELIER DREISEITL

- Avoid design features that stimulate stress such as: pointed or pierced forms, ambiguous or abstract art, reptilian-like tessellated scale patterns, snakes, spiders, and dark, cavelike spaces (Ulrich 1999; Ohman 1986).
- Provide opportunities to view wildlife by creating habitat and providing food, shelter, and water sources.
- Provide a focal point or positive distraction, such as sculpture, water feature, or unique vegetative specimen; research has shown that people—especially young children—prefer natural settings with water features (Ulrich 1993).
- Use landscape elements such as low walls, fences, vegetative screening, or topography to create a sense of enclosure that is both comfortable and safe.
- Encourage site users to explore the landscape more fully.
    - Provide multisensory experiences such as touching water, smelling and tasting vegetation, listening to birds, or feeling the warmth of the sun.
    - Signal ease of movement with design cues such as clear pathways and views of comfortable seating. Provide movable seating to allow site users to modify the environment to meet their needs.
    - Create comfortable outdoor microclimates that respond to the climatic conditions of the site and encourage year-round use. Examples include providing shade, windbreaks, or places to lounge in the sun.
    - Implement safety design strategies, such as those offered by CPTED or similar safety design guidelines.
- Mitigate noise pollution.
    - Conduct a noise-level study as part of the site inventory to determine existing equipment or areas that exceed the maximum acceptable noise level standards of 55 decibels.
    - Work with building architects and engineers to strategically locate and insulate HVAC machinery and other equipment.
    - Advocate for traffic-calming measures to reduce speed and alleviate noise.
    - Design the site to keep site users away from excessive noise.
    - Strategically locate outdoor noise barriers, such as dense foliage, earth berms, walls, or buildings. The most effective location for a noise barrier is very close to either the source or the receiver. Broad-leafed trees reduce noise better than conifers, and noise abatement is more effectual when the foliage extends to the ground (Bucur 2005). To effectively reduce noise, a dense band of vegetation at least 100 feet (30.48 m) wide is required.
    - Design in pleasurable sounds such as fountains or the rustling of leaves to provide a distraction and mask the objectionable noises (Thompson and Sorvig 2000).

Social Interaction

Humans have an inherent need for frequent social interaction and the development of stable and enduring relationships (Baumeister and Leary 1995). A lack of community interaction or feeling of belonging can not only cause emotional distress but also compromise the immune system and reduce life expectancy (Berkman and Syme 1979).

Neighborhoods with social ties are more likely to develop a strong sense of community that encourages residents, business owners, and other stakeholders to help one another, mobilize for community purposes, and defend their neighborhood against crime (Perkins 1990). Opportunities to build social networks are greater when people spend time outside their homes and businesses in green spaces such as front yards, neighborhood trails, and other common spaces or community parks and gardens. In addition to providing a gathering area, landscapes can also provide a setting where people are more relaxed and therefore willing to socialize (see Figure 3.7).

FRAUENWOHNEN

■ FIGURE 3.7 Residents of the FrauenWohnen in Munich gather around the courtyard well on a hot summer day.

Successfully designed sites can support a variety of impromptu and organized gatherings that benefit the health and well-being of site users and the community as a whole. Spaces should be specifically designed to meet the unique needs and desires of site users and other potential groups. Strategies for improving outdoor social interaction include the following:

- Create a variety of comfortable gathering spaces that feel safe and can accommodate different group sizes. When selecting locations, consider the microclimates of the site and potential for year-round use. Design spaces that can be easily seen from surrounding buildings or walkways. Minimize the use of visual obstacles and avoid other design features that provide a space for potential assailants to hide.
- Provide comfortable places for site users to sit and people-watch or socialize. Options may include movable furniture that allows groups to organize the space to best fit their needs or a variety of permanent seating options, such as stairs, seat walls, and benches.
- Locate gathering spaces near areas that are convenient and naturally attract activity, such as mailbox stations, food vendors, building entrances, or along major pedestrian routes.
- Provide a focal point or special item of interest, such as sculpture, a water fountain, or outdoor games, which can serve as a conversation starter and gathering spot.
- Design space for community activities that attract visitors, such as farmers' markets, festivals, or family gatherings.
- Provide amenities that attract visitors, such as electricity, free wireless service, or stages for performances.

Special Considerations for Children

Sites can serve a special and valuable purpose when they encourage children to interact with the natural environment. The diverse, interrelated, and dynamic components of nature stimulate children's innate curiosity to explore, experiment, and learn in multisensory ways that extend beyond what can be experienced indoors (Fjortoft 2004) (see Figure 3.8). As with adults, nature can improve a child's mental performance, reduce irritability, and expand socialization skills. Spending time in natural settings has also been shown to enhance children's attention spans and reduce symptoms of attention deficit hyperactivity disorder (Taylor and Frances 2009).

MILLER COMPANY LANDSCAPE ARCHITECTS

■ **FIGURE 3.8**
Formerly an asphalt lot, the Sherman Green Schoolyard allows students and teachers to extend the learning environment outdoors.

The American Academy of Pediatrics is an advocate of play and emphasizes its importance to healthy childhood development. Play has been shown to assist children in building creativity, imagination, and dexterity, as well as physical, cognitive, and emotional strength (Ginsburg 2007). Children often desire more complex, challenging, and exciting play environments than traditional playgrounds typically offer (Fjortoft 2004). Landscapes commonly used by children, particularly those in residential, neighborhood, and schoolyard settings, can reunite children with nature and help prepare them for the challenges of creating a sustainable society (see Figure 3.9).

SIEGFRIED J. GRAGNATO / © ATELIER DREISEITL

■ **FIGURE 3.9**
Access to water and interactive features are integrated throughout the stormwater management system of Arkadien Asperg.

Design strategies to immerse children in the outdoors include the following:

- Design landscapes that encourage children to play spontaneously without adult assistance. Locate play spaces in areas visible from within buildings, and outdoor seating areas to help adults become comfortable with children spending time outdoors independently.
- Encourage the independent mobility of children by providing neighborhood trails and linkages that are separate from vehicular traffic and safe for pedestrians and bicycles (Moore and Marcus 2008). Link parks, natural areas, and other spaces children may play in to the trail system.
- Use a mosaic of vegetation, natural materials, and varied topography to create diverse and dynamic landscapes that encourage versatile play and opportunities for exploration, creativity and improved motor fitness. Possibilities for play reflect the diversity and interest of the landscape itself (Fjortoft 2004). The exclusive availability of large playground equipment limits children's experiences and opportunities for well-rounded development (DeBord et al. 2003).
- Consider how vegetation and other landscape features will change throughout the seasons and what play opportunities these changes may provide. Seasonal changes offer new landscape forms—such as autumn leaves, bare branches, or snowy slopes—that provide different play habitats within the same setting (Fjortoft 2004).
- Select materials that are resilient and can tolerate active play. Protect the biological integrity of the site by considering the frequency of the use and ability of the landscapes to withstand the proposed activities (Fjortoft 2004).
- Provide a variety of natural and manufactured "loose parts" such as branches, pinecones, ropes, digging utensils, wheeled toys, and stones for games and building activities. Construction play is motivated by the excitement of the building process, not the end product. Often, when the construction of a project is finished it no longer holds interest for children, and a new project begins. Construction play encourages various forms of learning—planning, finding materials, fitting pieces together—that improve both cognitive processes and gross motor skills (Fjortoft 2004; DeBord et al. 2003).
- Avoid boring and monotonous designs that do not encourage creativity or provide a sense of challenge and excitement. Design landscapes that allow children to take safe risks while testing their emerging abilities. Safety issues should be addressed, but avoiding all risk is not the solution, as doing so limits children's sense of accomplishment and opportunity to master new skills and challenges (Little and Wyver 2008).
- Provide age-appropriate access to water that children can touch, manipulate, and play in. Explore opportunities for the water features to be safely incorporated into the site's stormwater management system—for example, capturing rainwater in structures that allow children to slowly release and direct the water in a sand or gravel play area.
- Create landscapes that attract wildlife and allow children to catch and release creatures such as fish, frogs, and insects.
- Provide adult-size seating and spaces in play areas to encourage adult/child interaction.

■ CASE STUDY

URBAN PLAY GARDEN

PROJECT TYPE: Single-family residential

LOCATION: San Francisco, California

SIZE: Wedge-shaped parcel 25 by 44 feet long (7.6 × 13.4 m)

Total of 1,100 square feet (102 m^2)

COMPLETION DATE: 2008

HIGHLIGHTED SUSTAINABLE PRACTICES:

Redevelopment of an underutilized urban site

Provides a safe and challenging environment for children to freely play outdoors, socialize, and interact with nature

THE SITE: The small urban lot in the Buena Vista Park neighborhood of San Francisco contains a three-story minimalist modern house and very little open land. Prior to redevelopment, the rear space was a very small-feeling, steeply sloped wedge-shaped parcel of land that was 25 feet at its widest point by 44 feet long (7.6 m × 13.4 m).

MARION BRENNER

■ **FIGURE 3.10**
Terraces and folding planes create a graphic urban garden when viewed from above, but at the garden level the focus is on adventurous play to draw children outdoors.

Design Overview

The minimalist modern architecture of the home is reflected in the Urban Play Garden, which promotes physical health and mental well-being through active interaction with the land (see Figure 3.10).

The primary goal of the project was to create a space where the client's children could play freely outdoors, something viewed as a basic human right and an important aspect of social sustainability. The site's topography allows the children and their friends to climb up a grass hill (using a rope) and race, roll, or slide down (on the concrete slide), the kind of thrilling adventurous play—the clients call the garden "a safe place for the children to feel bold"—considered essential for children's connection to the outdoors (see Figure 3.11). The garden's separation from the adult areas of the house gives the children a sense of their own place, and the availability of natural materials—sand, water, twigs, leaves, and flowers—to build with and support imaginative play. The bench that allows adults to enjoy or supervise the children's play is at the

continues

URBAN PLAY GARDEN *(CONTINUED)*

top of the garden, next to the house—at the greatest distance from the children's domain. Flat areas of the garden are used for ball games and to pitch a tent. The children participate in the stewardship of the landscape, where they dig, plant, and choose herbs, bulbs, and flowers to grow for tea parties and bouquets.

The design strategy was inspired by the work of Tadao Ando and his beautifully clean, smooth concrete walls and efficient use of space. The snap-tie concrete walls, colored to match the smooth stucco of the house, retain the terraces and planting beds and visually extend the architecture into the garden. The grading design balanced cut and fill, and the vegetation is drought tolerant and irrigated, only as necessary, by drip-irrigation. Water from the entire play garden is collected and directed to a bioretention drain. Once this area is fully saturated, then water flows into a stormwater drain.

MARION BRENNER

■ **FIGURE 3.11**
The upper and lower terraces can be explored by stairs, rope, or slide.

PROJECT TEAM

- **LANDSCAPE ARCHITECTS**
  Blasen Landscape Architecture
  Eric Blasen, Landscape Architect and Lead Designer
  Silvina Blasen, horticulturist
  Gary Rasmussen, project manager
  http://www.blasengardens.com/
- **ARCHITECT**
  Tim Gemmill
  http://www.gemmilldesign.com/
- **INTERIOR DESIGNER**
  Mark Cunningham
  http://www.markcunninghaminc.com
- **LANDSCAPE CONTRACTOR**
  Frank & Grossman
  http://frankandgrossman.com/
- **GENERAL CONTRACTOR**
  Creative Spaces
  http://frankandgrossman.com

REFERENCES

American Institute of Stress. 2010. Job stress: Absenteeism due to job stress has escalated. Retrieved May 5, 2010, from http://www.stress.org/job.htm.

Baumeister, R., and M. Leary. 1995. The need to belong: Desire for interpersonal attachments as a fundamental human motivation. *Psychological Bulletin* 117 (3): 497–529.

Berkman, L., and S. L. Syme. 1979. Social networks, host resistance, and mortality: A nine-year follow-up study of Alameda residents. *American Journal of Epidemiology* 109:186–204.

Bucur, V. 2005. Urban forest acoustics. Conference paper. Tecnico Acustica Terrassa. Champenoux, France.

Calorie Count. 2010. *Lawn & Garden*. Retrieved May 5, 2010, from http://caloriecount.about.com/activities-lawn-garden-ac8.

Centers for Disease Control. 2008. The benefits of physical activity. http://www.cdc.gov/nccdphp/dnpa/physical/everyone/health/index.htmhttp://www.cdc.gov/nccdphp/dnpa/physical/everyone/health/index.htm.

DeBord, K., R. Moore, L. Hestenes, N. Cosco, and J. McGinnis. 2003. *Making the most of outdoor time with preschool children*. Raleigh, NC: North Carolina Cooperative Extension Service.

Fjortoft, I. 2004. Landscape as playscape: The effects of natural environments on children's play and motor development. *Children, Youth and Environments* 14:21–44.

Ginsburg, K. R. 2007. The importance of play in promoting healthy child development and maintaining strong parent-child bonds. *Pediatrics* 119 (1): 182–91.

Kaplan, R. 1993. The role of nature in the context of the workplace. *Landscape and Urban Planning* 26:193–201.
The well-being of the workforce is researched here relative to the role of nature.

Kuo, F. E., and W. Sullivan. 2001. Aggression and violence in the inner city: Does vegetation reduce crime? *Environment and Behavior* 33 (3): 343–67.

Little, H., and S. Wyver. 2008. Outdoor play: Does avoiding the risks reduce the benefits? *Australian Journal of Early Childhood* 33 (2): 33–40.

Moore, R., and C. Cooper Marcus. 2008. Healthy Planet, Healthy Children: Designing Nature into the Daily Spaces of Childhood. In *Biophilic design*, S. R. Kellert, J. Heerwagen, and M. Mador, 153–204. Hoboken, NJ: John Wiley & Sons.

Ohman, A. 1986. Face the beast and fear the face: Animal and social fears as prototypes for evolutionary analyses of emotion. *Psychophysiology* 23:123–45.

Olshansky, S. J., et al. 2005. A potential decline in life expectancy in the United States in the 21st century. *New England Journal of Medicine* 352 (11): 1138–45.

Perkins, D. D., P. Florin, R. C. Rich, A. Wandersman, and D. M. Chavis. 1990. Participation and the social and physical environment of residential blocks: Crime and community context. *American Journal of Community Psychology* 181:83–115.

Taylor, A., and K. Frances. 2009. Children with attention deficits concentrate better after walk in the park. *Journal of Attention Disorders* 12:402–9.

Thompson, J. W., and K. Sorvig. 2000. *Sustainable landscape construction: A guide to green building outdoors*. Washington, DC: Island Press.

Ulrich, R. S. 1993. Biophilia, biophobia, and natural landscapes. In *The Biophilia Hypothesis*, S. Kellert and E. O. Wilson. Washington, DC: Island Press.

———. 1999. Effects of gardens on health outcomes: Theory and research. In *Healing gardens: Therapeutic benefits and design recommendations*, C. C. Marcus and M. Barnes, 27–86. New York: John Wiley & Sons.

World Health Organization. 2010. The definition of health. Retrieved June, 29, 2010, from https://apps.who.int/aboutwho/en/definition.html

VINCE STAMEY / BIGSTOCK.COM

FIGURE 4.1
Smoggy haze over Los Angeles.

CHAPTER 4

Sustainable Solutions: Air Pollution

AIR POLLUTION IS A CONSIDERABLE threat to the environment and to the health of billions of people worldwide. Over the last several decades, government policies and actions have reduced specific anthropogenic emissions and pollutant exposure; however, population growth and the subsequent increase in fossil fuel consumption continue to make air pollution a major concern.

Diminished air quality is commonly associated with developing countries; however, many cities in developed regions, including the United States and Europe, have air pollution levels that are unhealthy. Researchers have found that 58 percent of Americans live in areas where they are regularly exposed to air pollutants that pose both short- and long-term health risks (American Lung Association 2010). And in the European Union's twenty-seven member states, fine-particle air pollutants are associated with more than 348,000 premature deaths every year (European Environment Agency 2010).

Sustainable site development can improve local and regional air quality by reducing both the embodied and operating energy of a site. In addition to reducing emissions, vegetation—a key component of a sustainable site—also removes pollutants from the air, sequesters carbon, and provides the oxygen we all depend upon.

This chapter explores the relationship between air pollution and site development. Major pollutant sources and their impacts are discussed along with strategies for reducing embodied energy and creating favorable microclimates that benefit the site and surrounding area.

Air Pollution: The Cause

Air pollution alters the chemical composition of the atmosphere, and in doing so impacts human health and terrestrial and aquatic ecosystems. It can occur in many different forms—solid particles, liquids, and gases—and is generated from both human activities and natural events, such as volcanic eruptions and dust storms. Since the Industrial Revolution, air pollution levels have dramatically increased in severity and scale.

The primary driver of urban air pollution is the combustion of fossil fuels. Over 80 percent of the world's energy demands are met by coal, oil, or natural gas (Ngo and Natowitz 2009) that release harmful air pollutants, including carbon dioxide, nitrogen oxide, and particulate matter.

Fossil fuels are the primary energy source used to construct and maintain sites. In the United States, equipment such as lawnmowers, string trimmers, and leaf blowers contribute about 16 percent of hydrocarbon emissions and 21 percent of carbon monoxide emissions from mobile sources. And nonroad diesel engines, such as construction equipment, contribute about 44 percent of diesel particulate matter and 12 percent of total nitrogen oxide (NOx) emissions from mobile U.S. sources (U.S. EPA 2003).

In addition to the direct release of air pollutants, site developments that intensify urban heat islands also contribute to poor air quality due to increases in the demand for cooling energy in buildings and the acceleration of the formation of ground-level ozone and smog (see Table 4.1).

■ **TABLE 4.1**

Common Air Pollutants Generated from Site Development

POLLUTANT	DESCRIPTION	EXAMPLE SOURCES	HUMAN HEALTH IMPACTS	ENVIRONMENTAL IMPACTS
Carbon monoxide (CO)	Colorless, odorless, poisonous gas.	Incomplete combustion of fossil fuels, including emissions from small engines typically used for lawn and garden applications.	Reduces the delivery of oxygen to the body's organs and tissues.	Contributes to the formation of smog and ground-level ozone.
Carbon dioxide (CO_2)	One of the most abundant gases in the atmosphere, CO_2 is vital in plant and animal processes, such as photosynthesis and respiration; however, it is also a major greenhouse gas emission and by-product of fossil fuel combustion.	Combustion of fossil fuels, cement manufacturing, and pig iron and aluminum production.	Asphyxiation and other health impacts associated with climate change, such as water shortages, extreme heat, and flooding.	Greenhouse gas that contributes to climate change.
Nitrogen Oxides (NOx)	Group of highly reactive gases known as nitrogen oxides that includes NOx, N_2O, and others.	Fertilizers and combustion of fossil fuels from land-based nonroad diesel engines, such as construction equipment.	Inflammation of the airways and reduced lung function; cause of bronchitis, pneumonia, and lower resistance to respiratory infections.	Greenhouse gas that contributes to global climate change, acid rain, eutrophication, ground-level ozone, and fine-particle pollution.

POLLUTANT	DESCRIPTION	EXAMPLE SOURCES	HUMAN HEALTH IMPACTS	ENVIRONMENTAL IMPACTS
Particulate matter (PM)	A mixture of extremely small particles and liquid droplets that can be carried over long distances by the wind. PM is potentially one of the most harmful to human health.	Combustion of fossil fuels. Examples include nonroad diesel engines such as construction equipment, dust from construction sites, and wind-induced soil erosion.	Penetrates sensitive regions of the respiratory system. Short-term exposure can cause irregular heartbeat, decrease lung function, and aggravate asthma. Long-term exposure can lead to the development of heart or lung disease and premature mortality.	Reduces visibility and contributes to the formation of acid rain and smog. Changes the timing and location of traditional rainfall patterns.
Sulfur dioxides (SO_2)	One of a group of highly reactive gases known as sulfur oxides.	The combustion of fossil fuels that contain sulfur, such as diesel and coal. Examples include construction equipment and concrete manufacturers powered by coal. Also produced by volcanoes.	Difficulty breathing; may aggravate existing cardiovascular and respiratory disease.	Reduce visibility and contribute to the formation of acid rain and smog. Can stain, discolor, and deteriorate concrete, stone, textiles, and paints.
Volatile organic compounds (VOCs)	Gases from solids or liquids.	VOCs are emitted from diverse sources, including solvents, automobiles, construction equipment, fertilizers, and pesticides.	Eye, nose, and throat irritation; headaches, loss of coordination, nausea; damage to the liver, kidneys, and central nervous system. Some VOCs can cause cancer in animals and are suspected or known to cause cancer in humans.	Contribute to the formation of ground-level ozone.
Ground-level ozone (O_3)		Secondary pollutant formed from complex photochemical reactions following emissions of NOx and VOCs.	Decreases lung function, causes coughing and shortness of breath, aggravates asthma and other lung diseases.	Greenhouse gas that contributes to smog; damages leaves and disrupts photosynthesis.

VOC SOURCE: http://www.epa.gov/iaq/voc.html#Health%20Effects
EPA AIR POLLUTANTS SOURCE: http://www.epa.gov/air/urbanair/

Air Pollution: How It Affects Our Lives

Air pollution can degrade building materials, reduce the provision of ecosystem services, and significantly harm human health. Depending on the weather conditions, pollutants can remain trapped over cities for extended periods of time or be transported hundreds of miles by prevailing winds, resulting in damage to other regions.

Air pollution impacts include the following:

- Leaching of soil nutrients
- Reduced water quality
- Vegetation loss in both aquatic and terrestrial ecosystems
- Decreases in fish and other wildlife populations
- Climate change
- Corrosion of metal and deterioration of stone and concrete
- Short-term effects on human health, including irritation of the eyes and throat, shortness of breath, and increased respiratory infections
- Long-term effects on human health, including chronic heart and lung disease, cancer, neurological and developmental damage, and premature death

People who work or exercise outside face increased health risks. Some segments of the population, such as the elderly, children, and individuals with chronic respiratory conditions, are more vulnerable to the exposure to air pollutants.

Additional impacts occur when the pollutants mix with each other or with the basic components of the air to form a new pollutant. Common examples include the formation of smog, acid deposition, and greenhouse gases.

Smog and Ground Level Ozone

Many urban areas around the world, including Los Angeles, Mexico City, Tokyo, and Rome, often experience the haze and odor of smog. Smog is formed by a chemical reaction between sunlight and atmospheric particulates such as nitrogen oxides (NOx), carbon monoxide (CO), and volatile organic compounds (VOCs) (see Figure 4.2). The burning of fossil fuels in motor vehicles, power plants, and factories are primary contributors. Increases in atmospheric temperatures caused by urban heat islands accelerate the formation of smog. When winds are calm, smog can remain trapped over cities for extended periods of time. London experienced this tragic phenomenon in the winter of 1952, when extensive coal burning and a temperature inversion caused windless conditions that resulted in five days of intense smog and over four thousand deaths and one hundred thousand illnesses. In response, the UK created the first act of legislation to address air pollution in the world.

■ FIGURE 4.2 Smog is formed by a chemical reaction between sunlight and atmospheric particulates such as nitrogen oxides (NOx), carbon monoxide (CO), and volatile organic compounds (VOCs). The burning of fossil fuels in motor vehicles, power plants, and factories are primary contributors.

Ground-level ozone is a major component of smog; it should not be confused with the naturally occurring ozone layer of the stratosphere (10 to 30 miles above the earth's surface) that filters the sun's ultraviolet radiation. Cities around the world monitor ground-level ozone and regularly inform citizens of potential hazards. This diligence is due to a vast array of health and environmental issues associated with smog and ozone that include:

- Irritation of the respiratory system and reduced lung function
- Aggravated asthma and damage to the lining of the lungs
- Reduced visibility
- Disruption of photosynthesis, which reduces plant growth, increases susceptibility to stress, and decreases the provision of plant-related ecosystem services

Smog and ground-level ozone are often considered a regional issue because pollutants can drift 600 miles (1,000 km) or more, causing air quality concerns in areas well beyond the pollutant source (Baird 1999). One example of this is the Canada/United States border, where smog is transported in both directions and has prompted intergovernmental cooperation to reduce the generation and transport of air pollutants.

Acid Deposition

Acid deposition, commonly known as acid rain, is a broad term that describes solid particles and any form of precipitation that contains higher than normal amounts of nitric and sulfuric acids (Miller 1998). Acid deposition is created when sulfur dioxide and nitrogen oxides chemically react with water, oxygen, and other substances to form mild solutions of sulfuric and nitric acids (see Figure 4.3). In the United States, approximately two-thirds of all sulfur dioxide and one-quarter of all nitrogen oxide emissions are generated by electric power plants that rely on fossil fuels (U.S. EPA 2010a).

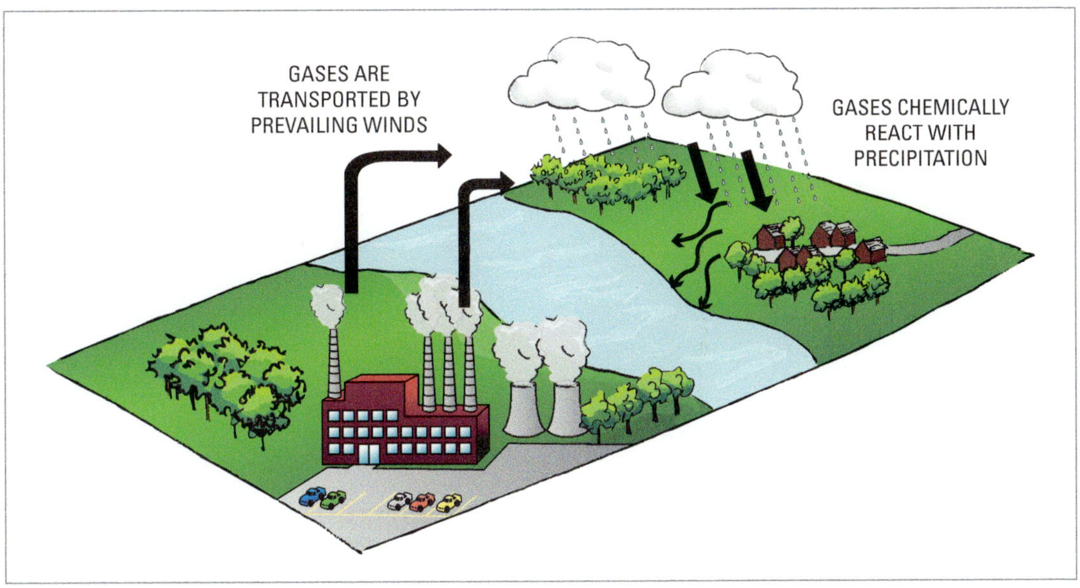

FIGURE 4.3 Acid deposition changes the chemistry of water bodies, harms habitat, damages buildings, and causes respiratory and cardiovascular health issues.

Acid deposition lowers the pH of soils and increases the aluminum levels of lakes, streams, wetlands, and other water bodies. This change in water chemistry harms vegetation and reduces the population, physical size, and biodiversity of aquatic life, primarily fish. Prior to falling to the earth, acid deposition also degrades visibility and harms human respiratory and cardiovascular health. In the United States, thousands of lakes and rivers have been affected by acidic deposition in the Northeast, upper Midwest, and mountainous areas of the western states. Acid deposition can also damage structures, such as buildings, monuments, and statues. Throughout the world, important cultural and historic sites, including the Taj Mahal, Roman Coliseum, and the Lincoln Memorial in Washington, DC, have been damaged by acid deposition. By damaging site structures, acid rain can lead to a short use life and the need for reconstruction, which magnifies the environmental and health impacts of the project.

Pollutants causing acid deposition can be transported hundreds of miles by prevailing winds, exporting the damage to other regions and countries. For example, most acid deposition that falls in Norway, Sweden, and the Netherlands originates from pollutants emitted in other European countries. In North America, pollutants from power plants in the Ohio Valley cause acid deposition in the eastern United States and southern Ontario, Canada (Baird 1999). It has been estimated by the Canadian government that fourteen thousand lakes in eastern Canada are acidic (U.S. EPA 2008).

Greenhouse Gases and Global Climate Change

Global climate change refers to significant and long-term changes in the earth's climate that are in addition to natural variability observed over comparable periods. The earth's climate has changed multiple times throughout history, with cycles of glaciation followed by warmer periods. Historically, natural factors, such as massive volcanic eruptions or slight variations in the earth's orbit have affected the global climate. However, in the late eighteenth century, human activities began to alter the composition of the earth's atmosphere and contribute to climate change.

Since the Industrial Revolution, global atmospheric concentrations of carbon dioxide have risen approximately 36 percent, principally due to the combustion of fossil fuels (IPCC 2007). Carbon dioxide and other greenhouse gases, such as methane (CH_4) and nitrous oxide (N_2O), trap heat and lead to atmospheric warming (see Figure 4.4).

■ **FIGURE 4.4** Carbon dioxide and other greenhouse gases trap heat and lead to warming of the atmosphere.

The rapid warming the earth is currently experiencing is unusual in the history of our planet and is occurring much more quickly than previous periods of climate change (NASA 2010). Elevated global temperatures have already begun to affect precipitation patterns, melt glaciers and ice sheets, and raise sea levels (IPCC 2007). Continued temperature increases are expected to lead to water shortages, inundate low-lying coastal areas (Karl et al. 2009), jeopardize agricultural production, and displace human populations.

Sustainable Site Strategies to Improve Air Quality

Sustainable site development not only reduces the generation of air pollutants but also cleanses the air, decreases atmospheric greenhouse gas concentrations, and produces oxygen. The most direct and effective ways for sites to improve air quality are to reduce energy use in all project phases, from materials manufacture to the ongoing operations and maintenance, and to sequester atmospheric carbon in soils and vegetation.

Strategies for improving air quality are numerous but often overlooked or underestimated. To address this issue, project teams should establish design goals and performance benchmarks for addressing air pollution at the outset of the project.

Site strategies to improve air quality include the following:

- Mitigate the urban heat island.
  - Reduce impervious surfaces.
  - Select high-albedo materials.
  - Shade heat-absorbing surfaces, such as roads, driveways, parking lots, and roofs.
- Reduce the embodied and operating energy of a site.
  - Select low–embodied energy materials.
  - Specify energy-efficient fixtures and equipment.
  - Utilize microclimatic design techniques that use vegetation and other site features to reduce the energy consumption of buildings.
  - Protect and incorporate existing native and other site-appropriate vegetation into the site design.
  - Strategically design a site to minimize maintenance practices that release harmful air pollutants.
- Sequester atmospheric carbon in soils and vegetation.

Vegetation and Air Quality

Vegetation provides numerous environmental, economic, and health benefits, making it one of the most important and obvious components of a sustainable site. Plants remove air pollutants, such as nitrogen dioxide, sulfur dioxide, ozone, carbon monoxide, and particulate matter of 10 microns (PM10) or less from the atmosphere by intercepting airborne particles and uptaking gaseous air pollutants through the leaf stomata and plant surface. In the coterminous United States, urban trees remove about 711,000 metric tons of air pollution per year, providing an estimated annual value to society of $3.8 billion (Nowak, Crane, and Stevens 2006) (see Figure 4.5).

Shade and evapotranspiration provided by plants also improve air quality by lowering air and surface temperatures, which in turn reduces the formation of ozone and emissions of such temperature-dependent pollutants as VOCs. Lowering surface temperatures is a particularly valuable in parking lots, or driveways, where shade from trees and other vegetation can reduce the evaporative emissions from vehicles.

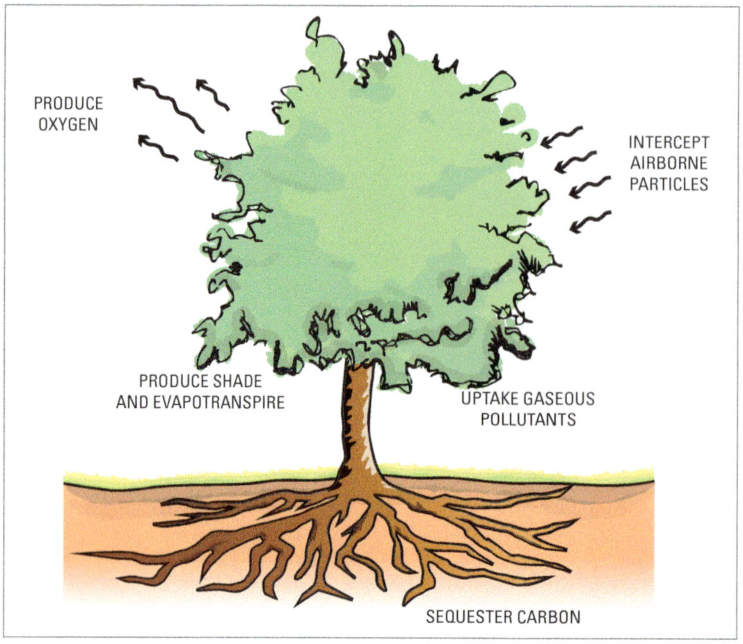

■ **FIGURE 4.5** Vegetation and air quality. Plants influence air quality in numerous ways: they uptake gaseous air pollutants, intercept airborne particles, release clean oxygen, sequester carbon, and provide shade and evapotranspiration, which lowers air and surface temperatures and reduces the formation of ozone and other temperature dependent pollutants.

Vegetation and certain tree species, including *Casuarina spp., Eucalyptus spp., Liquidambar spp., Nyssa spp., Populus spp., Quercus spp., Robinia spp.*, and *Salix spp.*, also emit VOCs to attract and repel insects. These natural VOCs, or biogenic emissions, are harmless until they react with nitrogen oxides (which are emitted by the combustion of fossil fuels) to form ground-level ozone and particulate matter. The overall benefits provided by trees are thought to outweigh any negative impacts of biogenic emissions; in areas with poor air quality, however, biogenic emissions are an important consideration when planning large-scale tree plantings.

■ RESOURCES

Hopper, L. J. 2007. *Landscape Architectural Graphic Standards: Student Edition*. Hoboken, NJ: John Wiley & Sons, pp. 52–53.

Urban Forest Ecosystems Institute. SelecTree: A Tree Selection Guide. "Biogenic emissions." http://selectree.calpoly.edu/biogenic.html.

Mitigate the Urban Heat Island

Landscape alterations brought on by urban development can generate and trap heat, resulting in changes in the local climate. This phenomenon, known as an urban heat island (UHI), describes cities or other developed areas that have warmer temperatures than their rural surroundings, forming an "island" of heat in the landscape (see Figure 4.6). Urban heat islands are the result of natural landscapes being replaced by dark and impervious surfaces, such as buildings and roads, which absorb solar radiation and release it as heat to surrounding materials and air masses. This is in contrast to vegetation, which uses solar energy to fuel the process of photosynthesis, during which moisture is released and cools the leaf surface and surrounding air.

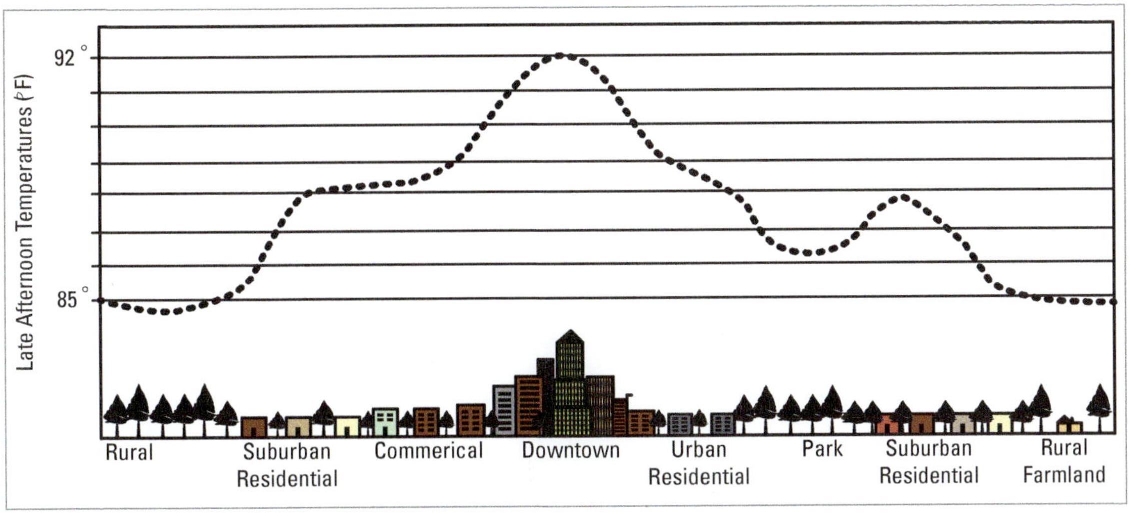

U.S. EPA

■ **FIGURE 4.6** Urban heat island profile.

Cities and suburbs impacted by urban heat islands can experience increases in air temperatures that are up to 10°F (5.6°C) warmer than the surrounding natural land cover. Temperature increases across a city are not consistent and depend largely on the land cover type, with parklands and lakes being cooler than adjacent impervious surfaces. Temperature increases can happen in any season, but differences are usually greater at night and most apparent when winds are weak. Urban heat islands have a range of effects that can impact cities and their inhabitants in the following ways:

- Increase energy consumption.
- Elevate air pollution levels and contribute to the formation of smog.
- Aggravate the risk of heat-related illness and mortality.
- Amplify uncomfortable outdoor summer conditions.
- Increase stormwater runoff temperatures, resulting in thermal water pollution and degraded aquatic habitat.

Through thoughtful design, project teams can create sites that mitigate the urban heat island while also providing other ecosystem services that benefit site users and the surrounding community.

Strategies include the following:

- Reduce impervious surfaces.
- Select high-albedo materials.
- Shade heat-absorbing surfaces, such as roads, driveways, parking lots, and roofs.

Reduce Impervious Surfaces

Impervious surfaces cover a significant portion of our urban environments. In addition to contributing to urban heat islands, impervious surfaces are also a cause of flooding and water pollution. A more detailed discussion of how to reduce impervious surfaces can be found in Chapter 5, "Sustainable Solutions: Urban Flooding and Water Pollution."

■ CASE STUDY

SAN FRANCISCO GREEN SCHOOLYARDS

PROJECT TYPE: Public schoolyards

LOCATION: San Francisco, California

SIZE: Between 4,000 square feet and approximately 15,000 square feet (372 m^2 and approximately 1,394 m^2)

COMPLETION DATE: 2007–2010

CLIENT: San Francisco Unified School District

HIGHLIGHTED SUSTAINABLE PRACTICES:

- Reduces impervious cover
- Mitigates the urban heat island
- Increases vegetative biomass
- Provides outdoor learning environments where students can interact with nature
- Renews a sense of community

THE SITES: Asphalt-covered schoolyards located at forty-five elementary schools and twelve middle and high schools.

MILLER COMPANY LANDSCAPE ARCHITECTS

■ **FIGURE 4.7**
This green schoolyard at Sherman Elementary replaces an asphalt lot with a natural learning environment that encourages play and thoughtful exploration.

Design Overview

A community-led effort to turn asphalt playground areas into sustainable outdoor learning environments is underway in the San Francisco public schools. The green schoolyard program was attached to two separate voter-supported public bond initiatives designed to address accessibility issues in the schools. Spearheaded by the San Francisco Green Schoolyard Alliance, a portion of the overall bond funding has been dedicated to the transformation of asphalt yards to foster higher academic achievement and increase environmental stewardship, creativity, and community building (see Figure 4.7).

Miller Company has worked closely with students, teachers, and parents in the design and development of the schoolyards, engaging each school in a barn-raising strategy that extends limited budgets. The engagement of local stakeholders in on-the-ground building efforts has also renewed a sense of shared community and pride for the schools.

Each garden creatively reflects the culture of the school and the unique design opportunities of the site. The schoolyard greening projects include outdoor classrooms and social spaces, rainwater harvesting, art, shade trees, riparian areas, and gardens of various themes (see Figure 4.8). The green schoolyards dovetail with newly expanded curricula, and it is apparent that positive changes are underway because of the new teaching environments.

SAN FRANCISCO GREEN SCHOOLYARDS *(CONTINUED)*

LAFAYETTE RAINWATER OUTDOOR CLASSROOM, MILLER COMPANY LANDSCAPE ARCHITECTS

■ **FIGURE 4.8**
Each garden creatively reflects the culture of the school and the unique design opportunities of the site. Active participation from students, teachers, and community volunteers in the development and ongoing maintenance of the schoolyard gardens has renewed a sense of community and increased support for the outdoor learning environments.

PROJECT TEAM

■ **LANDSCAPE ARCHITECTS**

Miller Company Landscape Architects

www.millercomp.com

Jeffrey Miller, Landscape Architect

Aaron Parr, Project Manager

Permeable Paving

Permeable pavements—also known as pervious or porous pavements—allow water and air to flow through the paving material (see Figure 4.9). Although initially designed to manage stormwater, permeable paving can also mitigate urban heat islands. The temperature and heat retention of the pavement is reduced by the evaporation of water within the pavement, and increased air and water flow through the pavement. When compared with conventional pavements, pervious paving provides more favorable growing conditions for tree and other plant roots. The improved conditions are due to the additional water and increased access to oxygen and nutrients in the underlying soil. A more detailed discussion of permeable paving can be found in Chapter 5, "Sustainable Solutions: Urban Flooding and Water Pollution."

■ RESOURCES

Calkins, M. 2009. *Materials for sustainable sites.* Hoboken, NJ: John Wiley & Sons.

Ferguson, B. K. 2005. *Porous pavements.* Boca Raton, FL: Taylor & Francis.

Low Impact Development Center: www.lowimpactdevelopment.org

Thompson, J. W., and K. Sorvig. 2000. *Sustainable landscape construction: A guide to green building outdoors.* Washington, DC: Island Press.

■ FIGURE 4.9
Vegetated permeable paving at the Turtle Creek Pump House was custom designed to accommodate vehicular traffic and allow site visitors to walk comfortably in heeled shoes.

High-Albedo Paving Materials

Solar reflectance, or albedo, is the measure of a material's ability to reflect sunlight. In general, albedo is associated with color, resulting in light-colored surfaces such as whites or pastels reflecting more sunlight than darker surfaces. Albedo is measured on a scale from 0 to 1, with 1 representing total reflectivity. A material with a value of 0.7 means that 70 percent of the solar energy hitting the surface is reflected and 30 percent is absorbed by the material. It is estimated that, for every 10 percent increase in the solar reflectance of pavement surface, temperatures can decrease by 7°F (4°C) (Pomerantz 2000).

In addition to albedo, it is also important to consider emittance, which is a material's ability to transfer and release heat. The Solar Reflectance Index (SRI) combines albedo and emittance into one measurement expressed as a fraction (from 0.0 to 1.0) or as a percentage (from 0 to 100). Materials with relatively high SRI values are often referred to as cool materials, such as cool roofs and cool pavements.

Light colors and smooth textures generally have higher SRI values. Some materials, such as new white portland cement (SRI of 86) and new grey concrete (SRI of 35) have general established values. Other materials may need to be tested using standards such as ASTM E1980, which defines SRI calculation methods. As a reference, LEED has established a performance target requiring an SRI of at least 29 for 50 percent of the paving area.

■ DESIGN CONSIDERATIONS

Sunlight reflected by high-albedo surfaces can create a glare that may limit visibility and is often uncomfortable to site users. Reflected energy can still contribute to urban heat islands if surrounding materials absorb it. In some climates, dark surfaces may be preferable for heating buildings, creating more comfortable outdoor microclimates, and melting snow or ice. In these situations, the winter benefits should be carefully weighed against the summer impacts, and options for shading the dark materials in the summer months should be explored.

High-albedo paving can help protect vegetation from extreme heat. Elevated air and soil temperatures from heat-absorbing walls and pavement can damage vegetative tissue and limit overall plant growth and performance. Researchers have found elevated rhizosphere temperatures to extend into the surrounding soil several feet beyond the pavement's edge, thereby limiting the soil volume available to plant roots (Celestian and Martin 2004).

SRI values can change over time as materials age and weather. For instance, asphalt tends to lighten as the binder oxidizes and the aggregate is exposed, and concrete typically darkens due to foot and vehicle traffic. Conventional maintenance practices such as the blacktopping of faded asphalt may be desirable for aesthetic purposes; it will, however darken the paving surface, resulting in lower SRI values. The maintenance and monitoring plan should recommend schedules and practices, such as removing dirt and oils or reapplying sealants, necessary to sustain the desired SRI value of materials.

■ RESOURCES

Calkins, M. 2009. *Materials for sustainable sites.* Hoboken, NJ: John Wiley & Sons.

U.S. EPA. 2005. *Reducing urban heat islands: Compendium of strategies: Cool pavements.*

Structural Soils

Trees are commonly planted adjacent to paving surfaces to provide shade, stormwater management, and other benefits. However, vegetated areas near conventional paving are often inhospitable environments for plants due to overly compacted soils, increased rhizosphere temperatures, and limited soil volume.

■ FIGURE 4.10 Structural soil underlying the Taylor 28 streetscape increases the rooting area available to the shade trees.

Structural soils are specialized mixtures of aggregate and soil formulated to support various pavement types while maintaining favorable growing conditions for vegetation. The specialized base course rooting media increases the volume of soil available to plant roots while maintaining the structural integrity of the pavement. Root volume and tree canopy size are positively related, making paved areas with adequate soil volume capable of growing larger, longer-lived, and more viable trees (Bassuk et al. 2005). As with porous paving, structural soils can provide stormwater management benefits. A more detailed discussion of structural soils and stormwater management can be found in Chapter 5, "Sustainable Solutions: Urban Flooding and Water Pollution."

■ RESOURCES

Ferguson, B. K. 2005. *Porous pavements.* Boca Raton, FL: Taylor and Francis.

Urban, J. 2008. *Up by roots: Healthy_structural soils and trees in the built environment.* Champaign, IL: International Society of Arboriculture.

Urban Horticulture Institute, Cornell University: CU Structural Soil. http://www.hort.cornell.edu/uhi/

Reduce the Embodied and Operating Energy of a Site

EMBODIED ENERGY

The sum of all the energy used during the life of a material or product, including the raw material extraction, manufacturing, transport, and disposal, is known as its embodied energy. The vast majority of energy is produced from the combustion of fossil fuels, which, as discussed earlier in this chapter, releases numerous air pollutants and greatly impacts human health and terrestrial and aquatic ecosystems. Because energy use and pollution are so closely related, the embodied energy of materials can be used as a strong indicator of materials that pollute (Thompson and Sorvig 2000). Many different variables come into question when evaluating and selecting materials for a sustainable site, and the relative importance of the environmental and human health impacts of a material is not always clear (Calkins 2009). Embodied energy analyses utilize a common component of all materials—energy use—to simplify the decision-making process.

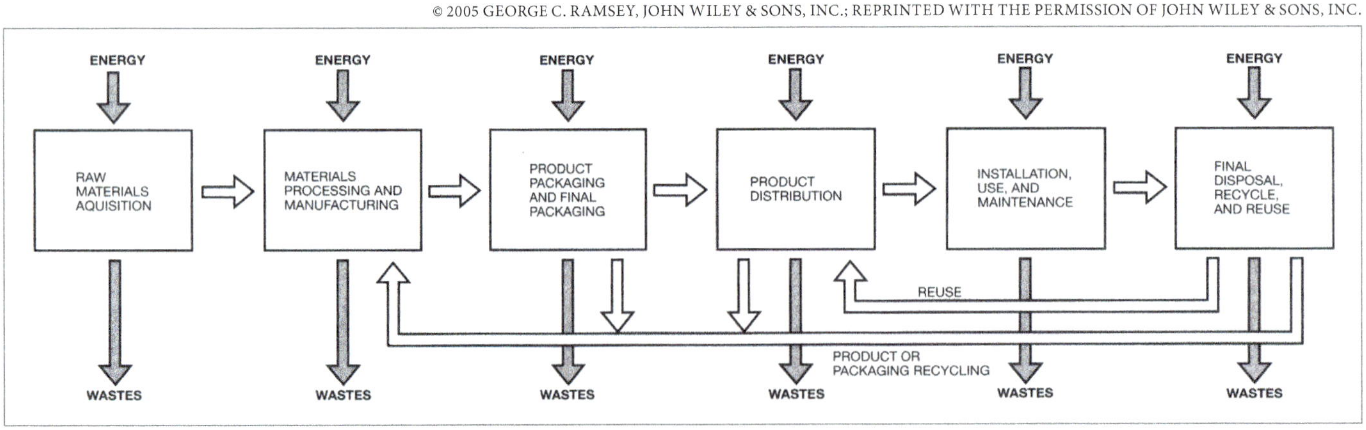

■ FIGURE 4.11
Life cycle of materials. Materials manufacturing is a consumptive and sometimes wasteful process. The typical life cycle of a material or product goes through a series of phases—raw material extraction, processing and manufacturing, packaging, distribution, installation, and disposal—with each step producing waste and requiring energy and resource inputs. Sustainable landscapes minimize the negative impacts of materials and products by altering the life cycle from a linear "cradle-to-grave" path to a cyclical "cradle-to-cradle" path, in which materials are not disposed of but rather reclaimed, reused, and recycled. Similar to natural ecosystems, a sustainable materials process turns waste into a resource. In doing so, habitat destruction and the harvesting of virgin materials to make new products can be avoided, and the release of harmful air, water, and soil pollutants prevented.

Products and materials with high embodied energy are those that have multiple manufacturing processes and include steel, copper, brick, synthetic nitrogen fertilizers, polyvinyl chloride (PVC), and high-density polyethylene (HDPE).

Embodied energy calculations are not an exact science; however, precise figures are not absolutely necessary for making environmentally sound decisions. When embodied energy estimates are not available, a material's relative embodied energy can be determined by its characteristics. In general, materials with low embodied energy skip or minimize energy consumption during one or several of the extraction, manufacturing, transport, and disposal phases. Common characteristics of low-embodied energy materials include:

- Reclaimed materials from the site or locations near the site
- Minimally processed materials such as stone pavers, aggregate, and wood
- Local materials that reduce transportation requirements
- Materials that use transportation sources that are less polluting—for example, trains use fuel more efficiently and have less emissions than gasoline or diesel trucks
- Materials that contain recycled content

The characteristics listed above are not equal and provide varying embodied energy savings. To reduce both embodied and operating energy, design teams should strive to select materials that have multiple low-embodied energy characteristics and require minimal energy to operate and maintain.

■ DESIGN CONSIDERATIONS

Embodied energy calculations do not directly address other human health or environmental impacts, such as the excessive generation of waste, water conservation during the production process, or the potential to reuse or recycle a product. These issues are better addressed using life-cycle analysis (LCA) or sustainability assessments (SA) tools, which are far more comprehensive and challenging evaluations (see Table 4.2).

■ TABLE 4.2

Analysis Tool Options

Many different variables, ranging from energy use to environmental and human health impacts, can be used to determine the most sustainable materials or product options for a site. Embodied energy analysis has been highlighted in this chapter because of its direct relationship to air pollution; however, it is but one of the many tools available. The following table contains a brief overview of the analysis tools that can be used to guide the material assessment and selection process.

ANALYSIS TOOL	DESCRIPTION	SCOPE	RESOURCES
Embodied energy (EE) analysis	The sum of all the energy used during the life of a material or product, including the raw material extraction, manufacturing, transport, and disposal, is known as the embodied energy.	EE studies determine energy use but are not sensitive to the specific energy source and its associated pollution levels. EE does not account for the health or ecological impacts of materials or products. Results will vary depending on the parameters of the study.	1. Calkins (2009) Materials for a Sustainable Site. 2. Athena Institute Impact Estimator for Buildings. 3. Cross and Spencer (2009) Sustainable Gardens.
Embodied carbon (EC)	The sum of all the CO_2 released during the life of a material or product.	EC studies typically correspond with EE figures unless the energy used to manufacture the material was "clean" energy that minimized the release of CO_2. EC does not account for the health or ecological impacts of materials or products and typically does not include other greenhouse gases. Results will vary depending on the parameters of the study.	1. Calkins 2009. 2. Athena Institute Impact Estimator for Buildings.

continues

■ TABLE 4.2

Analysis Tool Options *(continued)*

ANALYSIS TOOL	DESCRIPTION	SCOPE	RESOURCES
Life-cycle analysis (LCA)	Comprehensive assessment of a material's environmental impacts throughout its entire lifespan.	LCA studies have primarily been conducted for building products and whole assemblies. Information for landscape materials is limited. The studies are a comprehensive and time-consuming activity often developed by professional life-cycle analysts. The complexity and level of information gathered depends on the researcher and parameters of the study. Results vary depending on the weight given to each environmental impact.	1. Athena Institute Impact Estimator for Buildings. 2. National Institute of Standards and Technology, BEES (Building for Economic and Environmental Sustainability) software.
Sustainability assessment (SA)	A series of questions and instructions for collecting data regarding environmental and human health impacts of materials or products. SA questions are not intended to provide one correct answer but rather to identify major impacts, hazards, and opportunities in order to guide the material selection process.	A less scientific method than LCA. Information can be gathered from a variety of resources, including manufacturers, government agencies, and material safety data sheets (MSDS). The outcome depends upon the priorities of the client and project.	1. ASTM E2129: Standard practice for data collection for sustainability assessment of building products. 2. ASTM E2114: Standard terminology for sustainability relative to the performance of buildings. 3. Calkins 2009.

SOURCE: CALKINS 2009.

Considering the raw materials used in the manufacturing process and the source of energy used to fuel the process is also important. For example, some materials, such as plastics and synthetic fertilizers, are made from fossil fuels, which increase their embodied energy. And some high-embodied energy materials, such as aluminum, are commonly produced in facilities that use renewable energy sources (for aluminum in the United States, it is hydropower) in their manufacture, which reduces the air quality impacts but may cause other environmental damage.

HERBERT DREISEITL / © ATELIER DREISEITL

Reclaimed Materials

Reclaimed materials are those that have been salvaged and diverted from the waste stream for future reuse (see Figure 4.12). They can be reused in whole form or disassembled and adapted for new uses with minimal processing. Reuse is one of the most effective strategies for offsetting the initial environmental and human health impacts that result from the manufacture of materials or products. Giving a material a "second life" allows the majority of the material's life cycle to be bypassed, thereby conserving significant environmental resources. The most significant environmental impact of reclaimed materials is typically the energy used in transport, refinishing, and installation,

■ FIGURE 4.12

Reclaimed railroad rails form an undulating wall at the Tanner Springs Park. The rails connect the brownfield redevelopment to its previous industrial use and form an artful backdrop to the cleansing biotope and lower pond.

which can be minimized further if materials are salvaged and reused on-site (Hopper 2007). From the perspective of design and education, the reuse of on-site materials can also enrich the visitor experience by providing insight into the previous use and history of the site, as well as generate designs with unique meaning and detail.

Establishing performance targets and identifying reclaimed materials early in the design process will help facilitate reuse and reduce waste. A survey of all existing building and landscape materials and their potential for reuse should be conducted in the site inventory phase of a project. When determining reuse potential, allow the material to inspire the design and be open to using materials in new and innovative ways. Vegetation, stone, and soil that may need to be removed in the construction process but that can be salvaged and reused on-site or in nearby projects should be included in the site analysis. Also consider how materials that cannot be reused in whole form might be carefully deconstructed and creatively reused for a completely new purpose, or salvaged for reuse on another site.

Recycled Materials

Recycled materials are those that are collected, reprocessed, and used again to make a new product. They lessen the need for virgin feedstock and avoid sending useful materials to the landfill; however, significant energy and other resources are often required during the recycling process (Calkins 2009). Recycled materials should therefore be considered only after options to reduce or reuse materials have been fully explored.

Materials containing postconsumer recycled content—that is, materials that were once a consumer item and have been diverted from disposal—should be prioritized over preconsumer content, which comes from the manufacturing process, and can often be broken down and remade into similar or different materials. The U.S. Environmental Protection Agency's Comprehensive Procurement Guidelines recommend minimum postconsumer and total recycled content percentages for many landscape materials and products, such as playground equipment, bike racks, and park benches. A database of vendors who sell or distribute the products is also provided. Although the guidelines are intended for federal purchasing agencies, it is a useful resource for any project.

© MITHUN | JUAN HERNANDEZ

■ **FIGURE 4.13** To discourage waste, dumpster bins were not allowed on the Lopez Common Ground construction site. All waste from the project was separated into recyclables, trash, and "up for grabs." Each member of the construction team took a turn hauling the items to the recycling and waste facilities.

■ RESOURCES

BuildingGreen.com, GreenSpec: http://www.buildinggreen.com/menus/

CIWMB Recycled Content Product Directory: http://www.calrecycle.ca.gov/rcp/

RecyclingMarkets.net: http://www.recyclingmarkets.net/

U.S. Environmental Protection Agency: "Wastes—Resources Conservation—Comprehensive Procurement Guidelines": http://www.epa.gov/epawaste/conserve/tools/cpg/index.htm

■ CASE STUDY

TURTLE CREEK PUMP HOUSE

PROJECT TYPE: Commercial

LOCATION: Dallas, Texas

SIZE: 0.854 acre (0.35 hectare)

COMPLETION DATE: 2005

HIGHLIGHTED SUSTAINABLE PRACTICES:

- Reuse of existing site materials and structures
- Reduced impervious surface
- Increase vegetative biomass
- Rainwater harvesting and reuse
- Preservation of a unique cultural and historic place

■ **FIGURE 4.14**
Pump House site master plan.

SITE DESCRIPTION PRIOR TO THE DEVELOPMENT OF THIS PROJECT: The pumping station for the township of Highland Park, Texas, was built in 1915 to supply water for this growing Dallas suburb. In the 1950s, the parks department used the facility as its headquarters. Defunct since 1999, the building and grounds succumbed to time and vandals. Prior to redevelopment, the site was peppered with remnants of the old pumping infrastructure, including two in-ground masonry storage tanks, a wellhead, and iron water mains. The existing site was 90 percent impervious surface and had minimal vegetation on the bluff and perimeter of the property.

Design Overview

The Turtle Creek Pump House is the adapted reuse and reinvention of an abandoned water-pumping station and surrounding site, both considered to be at the end of their physical lives. It is a transformation of the industrial into the artistic, incorporating original mechanical equipment as well as elements of sustainable design. The site is currently used as a salon, a temporary flat, and a place for art and play.

Rather than see a piece of historic Dallas property sold for lot value, a neighbor adjacent to the old Turtle Creek Pumping Station bought the land with the express intent to redefine it. With respect for the site's original use, the client wanted not only to reclaim the old station and site but also to reinvigorate it with a new purpose—to lift the spirit by encouraging social interaction, intellectual discussion, and fun through imaginative uses of water (see Figure 4.14). The client also wanted the project to serve as a visual demonstration of environmental stewardship.

TURTLE CREEK PUMP HOUSE *(CONTINUED)*

Because the redefinition of the site was as important to the success of the project as the restoration of the buildings, the site was designed from the outside in. The landscape architects acted as both inventors and collaborators, working closely with the architect, the industrial reclamation programmer, and environmental artists to realize the client's goals.

Consistent with the desire to incorporate sustainable design elements, the team responded by designing a garden with intensively planted native Texas trees, grasses, and perennials (see Figure 4.15). A custom-designed vegetated porous paving system was installed for the parking lot and driveways and crushed granite was used in the motor court. This design was implemented to reduce runoff, allowing for the absorption of surface water into the ground. Rainwater from the roof was captured in gravel sumps to facilitate its reuse in the garden and avoid water being piped off-site. A buffalo sod (*Buchloe dactyloides)* green roof was also constructed atop a new bathroom facility to help minimize impervious cover and provide an overlook of the water tanks below (see Figure 4.16).

TOM JENKINS, DALLAS, TEXAS

■ **FIGURE 4.15**
Dry garden planted entirely with little bluestem (*Schizachyrium scoparium*), a native bunch grass. Penetration in the south tank wall opens to a sod green roof, which overlooks the tanks.

TOM JENKINS, DALLAS, TEXAS

■ **FIGURE 4.16**
Water rushes over the top of the south tank wall onto crushed granite rocks. Tank walls were water washed rather than sand-blasted to preserve the patina of the water line, and native plants were added gingerly in an existing void to soften the coldness of the old tank while preserving the sense of an empty vessel.

continues

TURTLE CREEK PUMP HOUSE *(CONTINUED)*

The design team salvaged and repurposed a large percentage from the existing site, diverting construction waste while showcasing the site's former life—everything from well covers to meter boards for benches to the concrete parking lot broken apart for stepping stones and courts (see Figure 4.17).

PROJECT TEAM

- **LANDSCAPE ARCHITECT**
  MESA
  Mary Ellen Cowan
  D.I.R.T. Studio
  Julie Bargmann
  Kate Orff
- **ARCHITECT**
  Cunningham Architects
- **INTERIOR DESIGN**
  Emily Summers Design Associates
- **GENERAL CONTRACTOR**
  Thomas S. Byrne

LISA MCCLEARY

■ **FIGURE 4.17**
The creative reuse of on-site materials welcomes visitors at the entrance garden, which includes the repurposed pump house, reclaimed concrete slab steeping stones, steel benches built from electrical panels, and a refashioned steel wellhead converted into a cocktail table.

Local or Indigenous Materials

Transportation methods, and the distance a product travels, are major considerations when determining the air quality impacts of a material. In the United States, transportation activities account for 28 percent of all greenhouse gas emissions, with trucks, ships, and trains making up 53 percent of the total inventory of gases and sinks (U.S. EPA 2007). Landscape materials are often heavy and bulky, making the distance that a product must be transported one of the most important considerations for site designers (Calkins 2011). As a reference, the Sustainable Sites Initiative recommends the following guidelines:

- Use only soils and aggregate that have been extracted, harvested, or recovered and manufactured within 50 miles of the project site.
- All other materials should be extracted, harvested, or recovered and manufactured within 500 miles of the project site.
- All growing facilities for vegetation should be located within 250 miles of the project site.

Researching regionally available materials early in the design process can help facilitate their use. Databases of local materials can be created and reused on future projects to help save time and money (Hopper 2007).

HEATHER VENHAUS

■ FIGURE 4.18 Cedar posts harvested on-site are used to create an arbor and vine tepee for the Little House Children's Garden at the Lady Bird Johnson Wildflower Center. Decomposed granite from a local quarry is used throughout the botanic garden as a trail surface and garden mulch.

Purchase New Materials That Can Be Reused or Recycled

The market for reclaimed and recycled materials is rapidly expanding; however, in some circumstances, purchasing new materials made from virgin feedstock may still be the only option. New materials do typically have high embodied energy, but by selecting materials likely to be reused or recycled, future resource use can be avoided. Reuse is one of the most effective strategies for offsetting the initial environmental and human health impacts of materials or products, and materials with recycled content reduce the need for virgin feedstock and the negative repercussions of raw material extraction and disposal.

Products or materials with the following characteristics increase their likelihood for reuse and recycling:

- Durability
- Modularity and/or standardized sizes
- Nontoxic
- Ability to be disassembled for reuse with reasonable effort and without damage that makes the material unusable
- Made of commonly and easily recycled materials such as metals, concrete, and brick
- Purchased from manufacturers with take-back programs

Durable goods may have higher embodied energy and require a larger initial investment, but they can also require less maintenance and have greater potential for reuse. The repeated replacement of low-quality or less-resilient materials will eventually outweigh any economic or environmental savings.

Design for Deconstruction

Project teams can help facilitate the reuse of materials by designing for deconstruction (DfD) and the intentional recovery of materials for reuse, remanufacturing, and recycling. The goal when DfD is to design site features in a way that allows them to be easily disassembled, thereby encouraging reuse or recycling. DfD helps to close the loop on construction waste and minimize energy consumption. Calkins (2009) outlines the following DfD principles and strategies:

- Establish DfD goals and performance benchmarks at the onset of the project.
- Develop a deconstruction plan that documents the materials and methods needed to successfully disassemble a structure or site feature.
- Specify materials that are durable, modular, and/or a standard size.
- Use simple and obvious connections that make the disassembly process easy and safe.
- Avoid connections such as mortar, adhesives, and welds that make a material difficult to disassemble and recycle. Use bolts, screws, and/or hand-nailed connections to ease disassembly and reduce the likelihood that a material will be damaged during deconstruction.
- Avoid such finishes as paint or sealers that can compromise the reuse or recyclability of the material due to additional cleaning costs.

■ RESOURCES

Calkins, M. 2009. *Materials for sustainable sites.* Hoboken, NJ: John Wiley & Sons.

Cross, R., and R. Spencer 2008. *Sustainable gardens.* Collingwood, Victoria, Australia: CSIRO.

National Institute of Standards and Technology. BEES online. http://www.nist.gov/index.html.

Operating Energy

The energy used in the day-to-day functioning of a site, including the energy required to heat and cool the buildings and to power outdoor lights, irrigation systems, and maintenance equipment, is known as its operating energy. A site's energy consumption and long-term operating costs are largely determined during the design process. Investigating a site's current and potential operating energy requirements early in the design process informs the analysis and research required to determine the most energy-efficient design solutions and integrate the buildings and landscape in a way that reduces overall energy consumption. Sites are uniquely positioned to reduce energy consumption with sustainable and relatively low-cost strategies that offer a multitude of economic, human health, and environmental benefits.

Strategies for reducing the operating energy of a site include:

- Employing microclimatic design techniques that use vegetation, materials, and other site features to reduce buildings' energy consumption
- Designing low-maintenance landscapes that strategically reduce site maintenance and the required use of land-care practices that release harmful air pollutants
- Specifying energy-efficient fixtures and equipment

How Site Design Affects Operating Energy

MICROCLIMATIC DESIGN: LANDSCAPES AND BUILDING ENERGY CONSUMPTION

Primarily due to their heating and cooling requirements, buildings are typically a site's largest energy consumers. Roughly 40 percent (22 percent residential and 18 percent commercial) of total U.S. energy consumption is used in buildings (U.S. EIA 2010). Figures are similar in other developed countries.

The energy consumption of buildings and the comfort of site users is largely a function of climate. A site's climate, however, is not uniform, but rather exists as a collection of microclimates—small, specific areas that differ from the broader regional climate.

Temperature, humidity, and wind speed will vary across a site due to factors such as plant structure, topography, and site materials (Brown and Gillespie 1995). People understand microclimates intuitively. On hot days, we seek breezy areas in the shade. In the city, the barefoot pedestrian hotfoots it across the street to the grass. On cold days, we look for sunny spots sheltered from the wind.

Strategic design and management of a site can create microclimates that reduce the energy consumption of buildings, mitigate the urban heat island, and improve the comfort of site users. In order to successfully manipulate microclimates, the site designer must first understand the regional climatic conditions, the influence of landscape elements on regional climate, and landscape design strategies that create comfortable microclimates for people and minimize the energy consumption of buildings (Brown and Gillespie 1995).

Determining the Microclimate of a Site

Detailed weather records can be provided for almost any region by local weather authorities or institutions such as the National Oceanic and Atmospheric Administration. Because this detailed level of information is not available for localized weather patterns at the site scale, design teams must look to surrounding landscape features to help them understand the site's unique microclimates. Below is an outline of natural and man-made factors that impact microclimate and assessment tools for determining their effects on site design.

SUN	To maximize benefits, designers must first understand the location of the sun throughout the year. A variety of methods can be used, including solar path diagrams, equations, and computer models (Hopper 2007). Shadow diagrams illustrate the location and length of shadows cast by objects in the landscape such as trees, buildings, and the surrounding geography. Several diagrams may be required to illustrate how shadows change throughout the day and at different times of the year. This information is useful not only for siting buildings but also for locating such features as outdoor patios, rest areas, and playgrounds that may require sun or shade in different seasons.
WIND	Wind, particularly in cold climates, can greatly affect a building's energy consumption and the comfort of site users. Wind speed and direction can be extremely variable, especially in urban settings, and is dependent on a variety of factors outside a site's parameters. Buildings and landscape features can modify the wind by redirecting it or changing its speed; however, they cannot completely stop the wind. Similar to the flow of water in a stream, the wind simply swirls around the object and continues to move across the landscape. Wind is difficult to accurately characterize, but careful observations of the site and surrounding landscape can help develop a mental image of what the wind is doing and how it is flowing (Hopper 2007). To visualize the impacts of wind, design teams should map prevailing wind conditions that exist throughout various times of the day and seasons, as well as elements on-site and in surrounding landscapes, such as buildings, walls, trees, and shrubs, that redirect or change wind speed (Brown and Gillespie 1995). Local weather stations can provide prevailing wind directions and wind rose diagrams can be used to illustrate seasonal patterns. In addition to observing the flow of wind across the site, wind measurement research from other landscapes with similar conditions can be used to determine potential wind patterns around standard landscape elements such as windbreaks and buildings (Hopper 2007).

continues

TOPOGRAPHY	The slope (the angle relative to horizontal) and the aspect (the compass direction a slope faces) are the two most influential microclimatic landform characteristics. Slope and aspect determine the amount of solar radiation a site receives and also influence local airflow. In the Northern Hemisphere, south- and west-facing slopes are typically hotter and drier than north- and east-facing slopes. Because cool air is denser than warm air, it naturally flows downhill and pools in drainage areas, valleys, or other low-lying areas in the landscape. Pockets of colder air can often be seen in warmer weather, as patches of light fog forming over river bottoms or wet vegetation. When the cold air drops below freezing, frost pockets will form and can cause damage to plants. Contour maps and field surveys can be used to understand the site's three-dimensional topography. Contour intervals should be sufficient to adequately capture variations in the topography and provide the information necessary for detailed site design.
VEGETATION	Extreme temperature fluctuations typically occur in areas with little or no vegetation, such as parking lots, conventional roofs, and deserts. Vegetation affects microclimates and moderates temperature by providing shade, increasing humidity, and modifying wind velocity and direction. The greatest microclimatic impacts are most likely to be gained from existing and mature vegetation, particularly trees. Projects that strategically incorporate existing vegetation into the site design can receive immediate benefits; major vegetation types—particularly evergreen and deciduous trees and shrubs—should therefore be mapped in the project's site inventory phase. Surveys should include the vegetation's location, species, general health, and estimated height.
BUILDING MATERIALS	Solar radiation is either absorbed or reflected by building and landscape materials. Energy that is absorbed heats the material and is slowly released, which in turn increases the surrounding air temperatures. The albedo, or solar reflectance, of a surface is generally associated with its color, with whites or pastels reflecting more solar radiation than darker surfaces. Porous pavements are cooled by water and air and typically have lower surface temperatures than conventional pavements. During the site inventory phase, all existing materials should be mapped and their effect on the microclimate assessed.

Strategies for Reducing the Operating Energy of Buildings

Climate and geographic location impact the ideal design of an energy-efficient site. Working with nature to develop comfortable microclimates is not a new idea; microclimatic design was standard practice throughout the world prior to the advent of climate-control technologies. The relatively low costs of energy and subsequent increase of mechanical heating and cooling systems have allowed the fundamentals of energy-efficient site design to be largely ignored—but not without significant environmental and human health consequences. Increases in air pollution, respiratory illness, and political strife have heightened awareness of the true cost of fossil fuels and provided new incentives for conservation. Site design and the strategic placement of buildings, vegetation, and other site features can improve the energy efficiency of buildings. Design strategies include the following:

- Orient buildings to use the sun's energy to collect and store heat in the winter, promote ventilation and cooling in the summer, and provide daylight throughout the year.
- Obstruct or channel wind flow to create favorable microclimates that reduce buildings' energy use.
- Strategically shade buildings to reduce solar heat gain and air-conditioner use in climates where overheating is a concern.

A fundamental strategy for improving the energy efficiency of buildings is passive solar design. Passive solar buildings capitalize on the sun's heat and light, greatly reducing—or in some cases, eliminating—the use of mechanical equipment and the need for external energy sources. Building orientation is a key component of passive solar design (see Figure 4.19). Properly oriented buildings use energy from the sun to collect and store heat in the winter, promote ventilation and cooling in the summer, and provide daylight throughout the year.

Building Orientation

Orienting the long axis of a building perpendicular to true north/south minimizes solar heat gain in the summer and maximizes solar benefits in the winter. Various site factors may make it necessary to shift the orientation somewhat; however, rotating a structure away from true south will reduce solar benefits. In addition, due to magnetic declination, compass readings may be inaccurate; based on the location of the site, specific adjustments may be needed to locate true north.

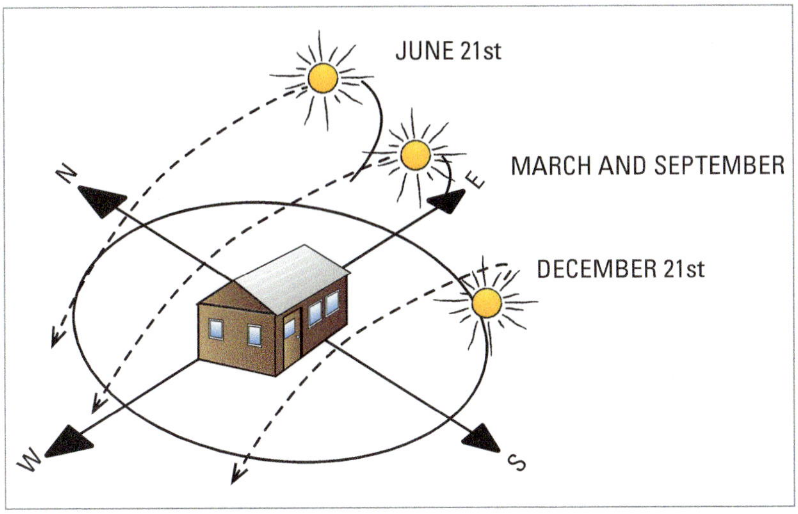

■ **FIGURE 4.19**
Ideal building orientation. The sun's path slowly shifts over the course of a year in a predictable pattern. In the Northern Hemisphere, during the winter months, the sun rises in the southeast and passes low through the southern sky, setting in the southwest. The sun reaches its lowest point on or around December 21, the winter solstice. In the spring and summer months, the sun passes more directly overhead and slowly shifts to the north, rising out of the northeast and setting in the northwest. On or around March and September 21 of each year, the spring and fall equinoxes, the sun is due east and west. The sun is highest in the sky on or around June 21, the summer solstice.

Understanding proper building orientation is also helpful when determining existing buildings' needs and energy-saving potential. Reorienting an existing building may not be an option; however, other design strategies using vegetation, materials, and wind can be used to create favorable microclimates that reduce the energy use of the building.

The most successful passive solar designs are those in which the building and landscape professionals work together to maximize the benefits provided by the living and built components of a site. The potential energy savings from proper building orientation can be lost if it does not work in conjunction with the site's vegetation and other natural conditions.

Obstruct or Channel Wind Flow

As wind moves across a landscape it reduces humidity, speeds evaporation, and increases the rate of air exchange between the inside and outside of a building (Reed 2010). Wind can be difficult to accurately characterize; however, vegetation and other site features can be used to obstruct or channel wind flow and reduce a buildings energy demand for heating and cooling.

In climates with cold winters, appropriately designed windbreaks can reduce a building's winter heating costs by approximately one-third (NREL 1995). Wind is often the most influential climatic element in winter months, particularly in buildings that are not well insulated or airtight (Brown and Gillespie 1995). Solar heat gain is typically not sufficient to overcome heat losses due to wind, making the obstruction of wind a top priority. The extent of protection a windbreak can provide is a function of its height and length. To maximize potential benefits, windbreaks should be oriented perpendicular to the prevailing winter winds and designed in a slightly convex shape that extends 10 to 20 feet (3 to 6 meters) beyond both sides of the object it is intended to protect (Reed 2010) (See Figure 4.20). There is no formula that fits every situation; in general, ideal wind protection is provided at distances two to five times the height of the windbreak (NREL 1995). When determining the best location for windbreaks, shadow diagrams should be used to ensure the windbreak does not interfere with the building's solar access. Special attention should be given to groups of buildings or trees, and gaps in windbreaks that may funnel and increase wind speed.

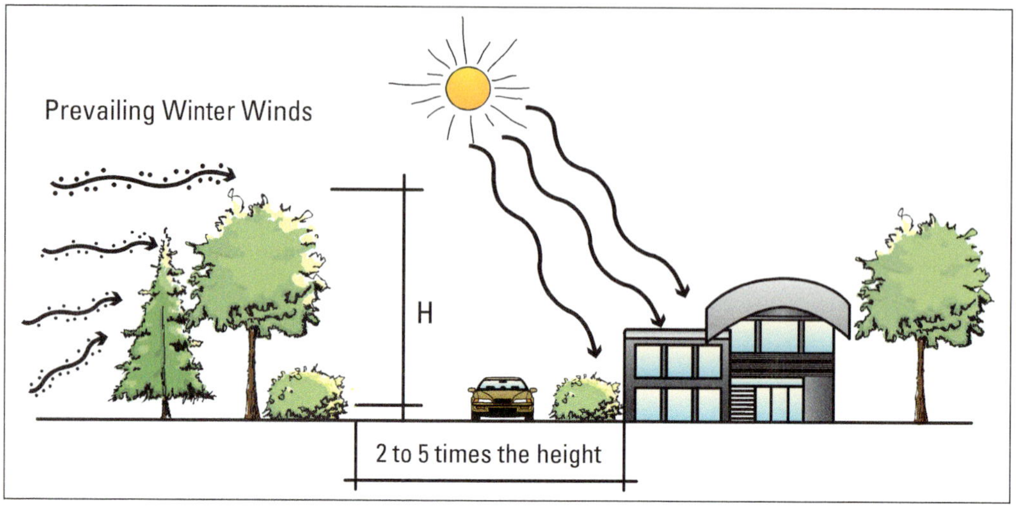

■ **FIGURE 4.20** Windbreak location. Ideal wind protection is provided at a distance two to five times the height of the windbreak. Special caution should be taken to ensure the windbreak does not shade the building in the winter months.

Windbreaks can be made from a variety of site features, such as vegetation, earth berms, buildings, and fences, and can combine a variety of materials and plant types. The most effective windbreaks impede the wind at multiple levels, extending from the ground to the treetops. Topography can influence wind direction and speed; however, unless the grade changes are significant, its effects are minimal. Using landforms in conjunction with other elements, such as trees or shrubs, will increase their efficacy. Trees are one of the most effective landscape elements for modifying wind speed and direction. A standard multiple-row windbreak consists of a windward (upwind) row of dense conifer trees or shrubs, interior rows of tall broadleaf trees, and leeward (downwind) rows of broadleaf shrubs or conifers. Windbreaks made up from a diverse suite of species will improve habitat conditions and reduce the risk of insect and disease.

As a general rule, a dense windbreak, such as a brick wall, has a greater effect on wind speed but protects a smaller area. Conversely, looser and more porous windbreaks, such as vegetation, have less of an effect on wind speed but impact a larger area (Brown and Gillespie 1995). This is due to low pressure behind the windbreak pulling air down and creating turbulence. Windbreaks that are less dense and allow air to pass through them moderate the low pressure and turbulence, thereby increasing the length of the downwind protected area (see Figure 4.21).

In addition to windbreaks located upwind, shrubs and vines can also be planted next to buildings to create dead airspace to provide insulation in winter and summer. This may be a more practical solution for small sites that cannot provide adequate distance between larger windbreaks and buildings. For solar heat gain purposes, sunlight needs to reach south-facing walls and windows. This strategy may be counter-productive in climates with hot, humid summers because of the increase in humidity, albeit slight, created by the vegetation. In this circumstance, the heating and cooling benefits must be weighed, and priority should be given to the planting strategy that offers the greatest energy savings.

Vegetation, topography, and other site features may also be used to cool a building by providing shade and directing wind flow. Breezes can be directed toward buildings by leaving an open channel in the landscape in the direction of the prevailing summer winds. Large gaps in windbreaks or buildings can also channel the wind and increase its speed. Summer breezes should only be directed toward a building if the structure uses natural cooling practices and is not mechanically air-conditioned.

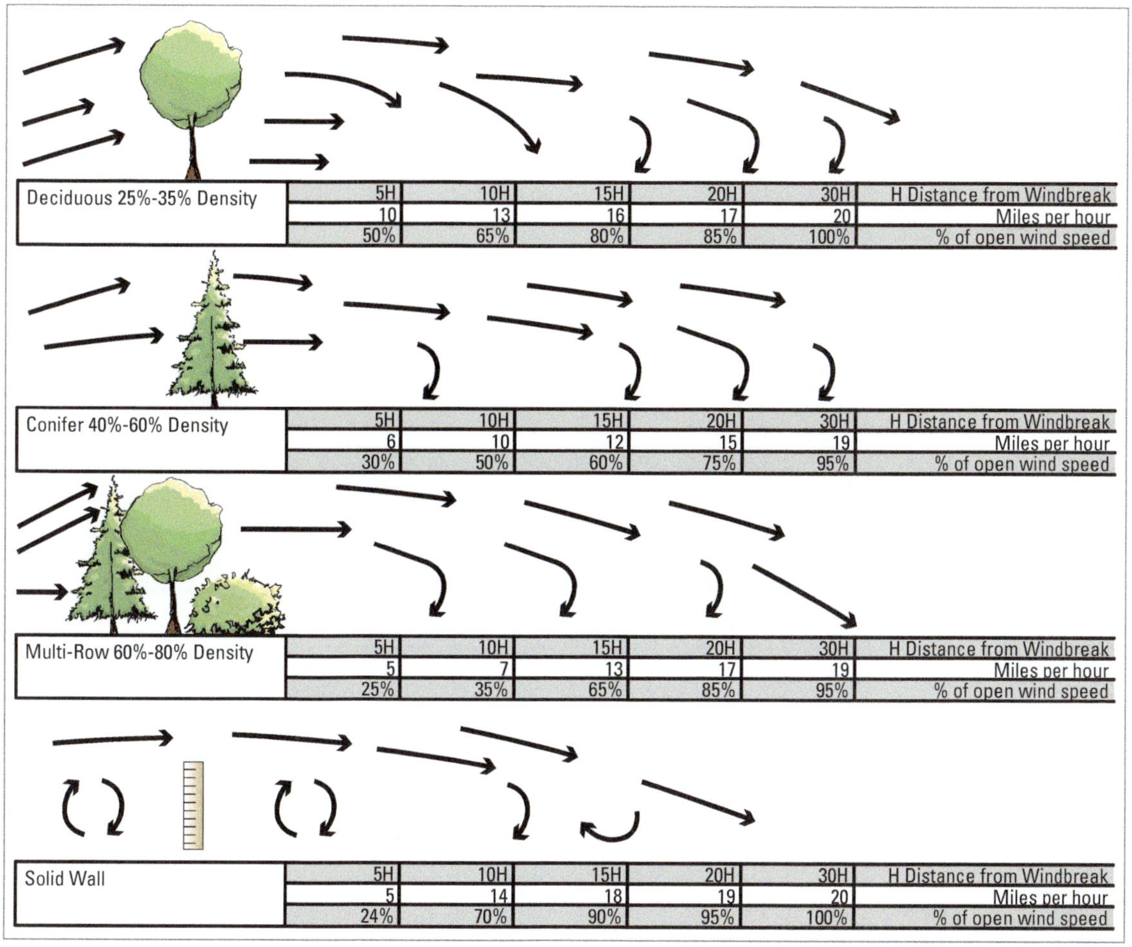

Windbreak		5H	10H	15H	20H	30H
Deciduous 25%-35% Density	Miles per hour	10	13	16	17	20
	% of open wind speed	50%	65%	80%	85%	100%
Conifer 40%-60% Density	Miles per hour	6	10	12	15	19
	% of open wind speed	30%	50%	60%	75%	95%
Multi-Row 60%-80% Density	Miles per hour	5	7	13	17	19
	% of open wind speed	25%	35%	65%	85%	95%
Solid Wall	Miles per hour	5	14	18	19	20
	% of open wind speed	24%	70%	90%	95%	100%

H Distance from Windbreak

SOURCE: U.S. NATIONAL RESOURCE CONSERVATION SERVICE

■ **FIGURE 4.21**
Wind-speed reductions of windbreaks with varying densities. The open wind speed in this example is 20 mph. H = the height of the windbreak.

Cool air is denser than warm air and naturally migrates to lower elevations in the landscape. Vegetation, earth berms, and walls can divert or direct cool, dense air as it flows downhill. These "frost dams" can be used to reduce or increase cold-air ponding around a building or landscape. See Figure 4.22 for an illustration of cold-air drainage and frost dams.

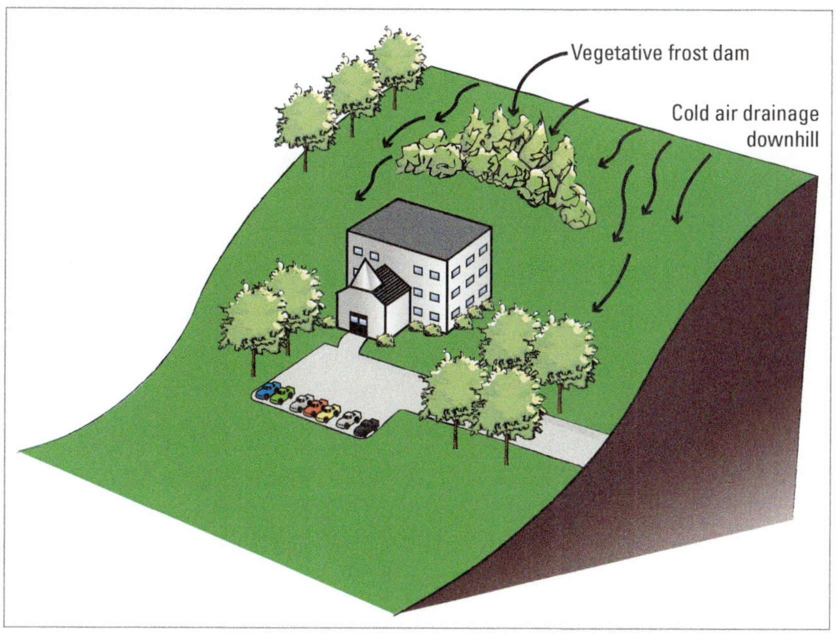

■ **FIGURE 4.22**
Frost dams. Cold air flowing downhill can be diverted or even halted by landscape features such as low walls, thick hedges, earth berms, or dense stands of trees.

CASE STUDY

LOPEZ COMMON GROUND

PROJECT TYPE: Multifamily development

LOCATION: Lopez Island, Washington

SIZE: Eleven single-family homes, two single-room-occupancy (SRO) program apartments, and land trust offices. The development disturbs just 1.6 acres (0.64 hec) of a 6.5 acre (2.6 hec) parcel.

COMPLETION DATE: 2009

CLIENT: Lopez Community Land Trust

HIGHLIGHTED SUSTAINABLE PRACTICES:

- Integrated design process
- Net-zero energy consumption
- Rainwater harvesting and reuse
- Food production
- On-site stormwater management and reuse

FIGURE 4.23
Homes are nestled into the base of the forested edge and are oriented due south to maximize solar access.

SITE DESCRIPTION PRIOR TO THE DEVELOPMENT OF THIS PROJECT: Located on Lopez Island in the San Juan Islands, the 6.5-acre (2.6 hec) site gently slopes to the southeast and was split into three distinct zones: meadow, transitional edge, and third-growth forest of both deciduous and conifer trees. The site is 3⁄10 mile (0.5 km) north of Lopez Village and is located within the urban growth area as designated by San Juan County. One-third of this previously undeveloped site was used primarily for hay production, while the rest remained forested. A mandatory 80-foot (24 m) buffer at the site's south end separates Lopez Common Ground from the adjoining 8-acre (3.2 hec) Fisherman Bay water treatment plant. The average rainfall is 25 inches (64 cm) per year.

Project Description

Lopez Common Ground is a mixed-income, affordable community for families earning less than 95 percent of the area income. The development accommodates a variety of uses, including housing, work, and agriculture, and aims to promote local self-sufficiency while preserving the rural character of the site. Intended to be a demonstration project, the development has strong documentation, which facilitates easy adoption by other developments.

The community of homes can be accessed by pedestrians and vehicles directly from the street and is a mere six-minute walk from the heart of Lopez Village. Rigorous siting exercises were conducted to minimize impact to the site. The preliminary site plan supported the Land Trust's goal of low-impact development by clustering housing, minimizing impervious surfaces, and following the land's natural form.

All buildings are oriented to maximize solar exposure for passive heat and light gain, as well as for active solar hot water and passive photovoltaic power generation. The community aims to achieve net-zero energy consumption on an annual basis through the use of passive solar design principles and on-site, renewable energy sources (see Figure 4.23).

LOPEZ COMMON GROUND *(CONTINUED)*

A central green space provides a communal "living room," with individual garden plots where residents can grow their own food or create flower gardens for relaxation and enjoyment. An aboveground tank collects rainwater on-site, which supplies water for toilets, washing machines, and irrigation. Residents are given an annual rainwater allotment, for which they are assessed charges only if their usage exceeds the allowance. A bioswale weaves through the site, transporting stormwater to a pond at the lowest point. The pond is slowly released to the wetlands south of the site or used for supplementary irrigation during the summer. Bioretention cells located along the lower edge of the parking lot catch surface water runoff, which is then infiltrated back into the soil. The plant palette reflects the site's ecology, and plantings were incorporated to support habitat, increase biodiversity, decrease water use, and provide functional and edible landscapes.

The design process included a number of integrated charrettes with residents, community stakeholders, and the Land Trust. Each meeting focused on identifying specific goals and milestones necessary for creating a sustainable, self-sufficient community.

By integrating passive solar design techniques and low-impact development strategies with solar power, Lopez Common Ground is already close to achieving net-zero energy and water use in far less than the five years originally projected (see Figures 4.24 and 4.25). During the first year of operation, solar production on-site was approximately 4,000 kilowatt hours per home, while the average use is 5,700 kilowatt hours per year; some homes actually produced more energy than they used annually. This resulted in a savings of over $550 annually per household for energy alone. Each year, residents are refunded approximately $750 thanks to Washington State's solar power incentive program.

■ **FIGURE 4.24**
Zero net water strategies for the community.

continues

LOPEZ COMMON GROUND *(CONTINUED)*

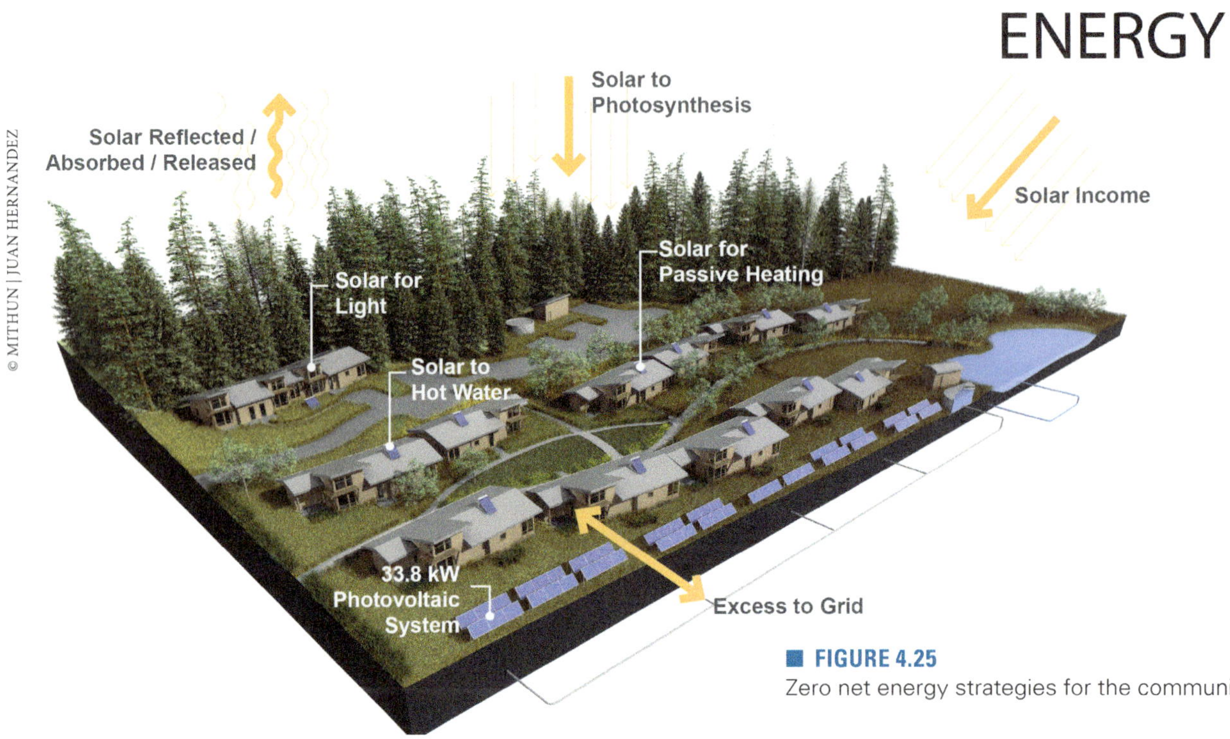

■ **FIGURE 4.25**
Zero net energy strategies for the community.

PROJECT TEAM

■ MITHUN

Bill Kreager, Architect
Bruce Williams, Architect
Tammie Schacher, Architect
Brian Cloward, Architect
Mike Fowler, Architect
John Fleming, Architect
Steve Cox, Architect
Tyrone Jordan-Oliver, Architect
Theo Manning, Architect
Erin Jacobs, Landscape Architect
Rob Matthews, Planner
Chuck Weldy, Specifications Writer

Chris Webb, Webb & Associates, stormwater strategies
John Hart, Hart Pacific Engineering
Stephen Yu, Yu and Trochalakis, Structural Engineers
Joe Bullock, ReSources, permaculture
Richard Hobbs, Strategy Design, Inc.
Israel Gaphni, Sound Mechanical Consulting
Dana Brandt, Ecotech Energy Systems

Shade

Strategically shading a building is one of the most cost-effective strategies for reducing solar heat gain and air conditioner use in climates where overheating is a concern (see Figure 4.26). Well-shaded buildings can have indoor air temperatures 8.7°F (4.8°C) to 20°F (1°C) cooler than similar non-shaded buildings (McPherson 1984), providing significant energy savings.

Shade improves the energy efficiency of a building by lowering surface temperatures of the roof and walls and reducing the transfer of heat from the outside air into the interior spaces. The energy efficiency of both new and existing buildings can be improved by shade; however, the following five factors must first be determined:

1. Months of the year and hours of the day that shade provides the greatest cooling benefits
2. Building surfaces and landscape materials that experience the greatest heat gains
3. The location of the sun when shading is desired
4. The desirable characteristics of plants or shade structures that best meet the site's needs
5. Where to locate vegetation or shade structures to provide the greatest benefits

Haphazardly removing or planting vegetation without understanding these important design factors can result in greater energy costs, rather than savings. A common example for cold weather climates is the location of trees or other shade structures on the south side of buildings. In order for buildings to benefit from solar heat gain, sunlight should reach as much of the south-facing wall (or north-facing wall in the Southern Hemisphere) and roof as possible between the hours of 9 am and 3 pm solar time. Locating trees to provide shade during summer months without considering winter shade patterns can result in the south-facing wall receiving shade during critical times of the day and an overall increase in energy consumption.

Sun and shadow diagrams illustrating the location and length of shadows from trees, buildings, and other site features can provide the design team with the information needed to take advantage of existing site conditions throughout the year. In addition to diagramming features on-site, the shade impacts from surrounding properties must also be considered, particularly if the adjacent area is at a higher elevation or properties are small or close together. Understanding a site's sun and shade patterns is useful for locating new buildings as well as for determining how site features impact the energy efficiency of existing buildings.

BILL TIMMERMAN

BILL TIMMERMAN

■ **FIGURE 4.26**
A vertical scrim vegetated with native vines helps to shade and cool a university building outside the Underwood Family Sonoran Landscape Laboratory.

A structural analysis of existing buildings can be used to determine the portions of the building where the greatest heat gains occur. Such an analysis typically includes items such as building orientation, window location, solar access, and construction materials and methods. Based on these findings, designers can then prioritize shading needs and select the most appropriate and effective design strategies.

During the summer months, the altitude of the sun shifts slightly to the north, making the building surfaces that receive the most sunlight the roof and east- and west-facing walls. Overheating early in the day can be prevented by shading east- and southeast-facing building surfaces; shade on the west and southwest walls and roof can significantly reduce peak indoor air temperatures and accelerate cooling in the afternoon and evening. Heavily shaded east- and west-facing walls provide greater energy savings for cooling than a heavily shaded roof, because the walls have longer periods of exposure to solar radiation and more favorable angles of incidence during the majority of the day (McPherson 1984).

A building gains substantially more heat through windows than insulated walls, making the shading of windows a priority in the summer. Shading air conditioner units can increase the efficiency of the system by as much as 10 percent (NREL 1995). To maximize efficiency, it is important to allow adequate airflow to and from the air-conditioning unit and provide continuous shade throughout the cooling season.

Because building surfaces are heated by both solar radiation and ambient air temperature, it is important to not only shade the building but also heat sinks in the surrounding landscape. Low-albedo landscape materials and impervious surfaces, such as asphalt roads, parking lots, and driveways, absorb and radiate significant amounts of heat. Shading these surfaces, particularly during summer months, is an important urban heat island mitigation strategy that will benefit not only the site but also the surrounding area.

PLANT SELECTION

Many landscape and architectural features can be used to provide shade; however, vegetation when used in mass provides the additional cooling benefit of evapotranspiration, which lowers ambient air temperatures. The combination of shade and evapotranspiration in densely vegetated settings has been shown to reduce outdoor air temperatures as much as 9°F (5°C) (U.S. Department of Energy 2010). When considering the shading potential of vegetation and making choices as to which plants should remain or be added to a site, bear in mind the following characteristics:

- **Foliation period:** The average period a plant is in leaf should align as closely as possible with the cooling requirements of buildings. Leaf seasons are averages, and microclimate conditions and maintenance practices such as irrigation and pruning can affect the foliation period.
- **Shade density:** The density of shade cast by vegetation is determined by the characteristics of the leaves and branches, and can be described as light, moderate, or heavy. For example, evergreen trees cast a heavy shade, allowing very little sunlight to pass through the canopy, while deciduous trees with small leaves and open canopies cast a light shade, allowing patches of sunlight to pass through. The cooling benefits provided by a plant increase with its shade density.
- **Mature size and form:** Shadow pattern and shade area are directly related to the size and form of the vegetation. Shade area is typically determined using the estimated size of mature vegetation; however, it may take some time before plants, particularly trees, to reach their full shade potential. To provide shade until vegetation is mature, design teams can use vines, shrubs, and ornamental grasses, which mature faster than trees, require less room to grow, and are less likely to damage

the building's foundation. Technologies such as green roofs and green walls can also be used to gain more immediate shade benefits.

- **Growth rate and life span:** When selecting vegetation, consider the plant's growth rate, hardiness, and estimated life span. Trees that grow rapidly will provide shade sooner, but most are short-lived and have weak wood that can be a hazard to site users and the property. A combination of fast- and slow-growing trees can be strategically located to provide shade in both the short and long term; the fast-growing species can be systematically removed as they become a hazard or are no longer needed. This kind of successional planting provides a high level of shading efficiency in a relatively short timeframe and sustains the shade for a long period of time without the hazards or high maintenance costs commonly associated with fast-growing tree species or the high initial cost and potential loss of planting large trees (McPherson 1984).
- **Distance between the ground surface and base of the canopy:** When the sun is low, sunlight passes through the gap between the ground surface and the lowest tree limbs. The branching height of trees can be carefully selected to block summer sun but allow lower-angle winter sun to reach buildings.
- **Maintenance requirements:** Design teams should select vegetation whose natural growth patterns and size are appropriate for the shade requirements of the site and does not require extensive maintenance.

Deciduous plants are useful in situations where both solar access and shade are needed during different times of the year. However, bare branches, stems, and trunks still cast significant shade during the winter. Evergreen vegetation provides shade year-round and is commonly used to slow and redirect the wind.

Vegetation planted close to a building will shade more area for longer periods of the day than plants of the same size located at farther distances. In dense urban areas or small sites where space is limited, it can be difficult to provide adequate room for vegetation. Support structures and technologies, such as green roofs, green walls, and structural soils, can make dual use of limited space and accommodate vegetation in tight spaces or in unconventional settings.

■ RESOURCES

Dines, N., and K. Brown. 2001. *Landscape architect's portable handbook.* New York, NY: McGraw-Hill.

McPherson, G. 1984. "Solar-control planting design." Washington, DC: American Society of Landscape Architects.

Reed, S. (2010). *Energy-wise landscape design: A new approach for your home and garden.* Canada: New Society Publishers.

Weather Data Depot. Heating and cooling degree-day reports. http://www.weatherdatadepot.com/#.

Green Roofs

Green roofs (see Figure 4.27), also known as vegetative, living, or eco-roofs, are specialized roofing systems that support vegetation on both sloped and flat roofing surfaces. They are comprised of a series of layers that, at a minimum, include vegetation, soil, or growing media and a waterproofing membrane. Some systems also require filter cloth, a drainage layer, and/or a root barrier. The continuous or near-continuous use of vegetation and other layers distinguishes green roofs from the conventional roof garden, which utilizes freestanding containers or pots on an accessible rooftop or deck.

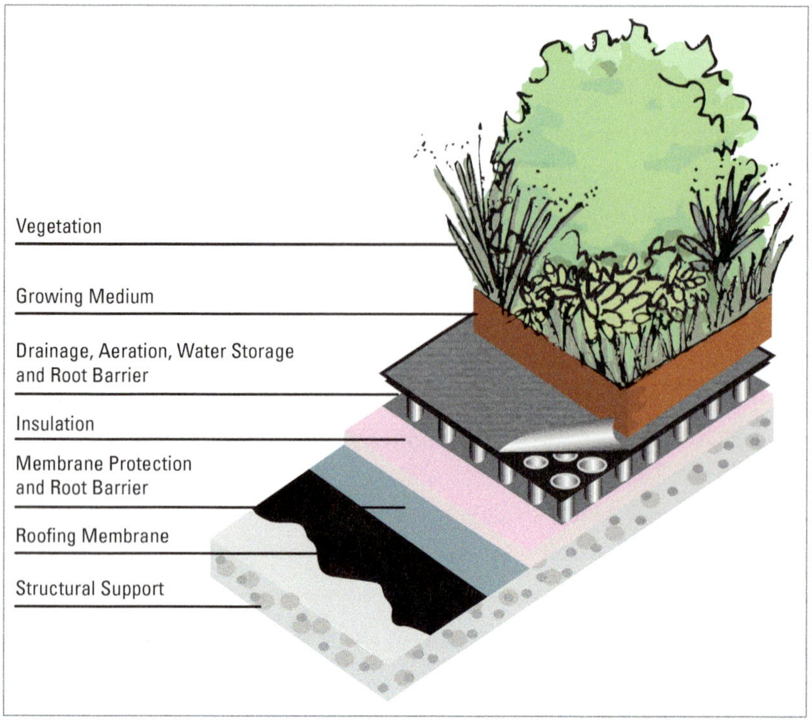

■ FIGURE 4.27 Green roof.

Green roofing systems are separated into two main categories: extensive and intensive (see Table 4.3). Extensive green roofs are relatively lightweight systems with a shallow growing media of 6 inches (15 cm) or less. They are designed to require minimal maintenance, and access is typically limited to maintenance personnel. Intensive green roofs are heavier systems that contain several feet of soil or specialized growing media and can support a wide variety of plant materials, landscape amenities, and outdoor uses. The vegetative roof type is dependent upon the loading capacity of the structure and the slope of the roof. Some green roofs are designed with features of both and are referred to as semi-intensive.

■ TABLE 4.3 Extensive and Intensive Green Roof Comparison

CHARACTERISTIC	EXTENSIVE ROOF	INTENSIVE ROOF
ACCESSIBILITY	Typically limited to maintenance personnel	Typical of a standard garden, intensive green roofs can support a variety of uses and landscape amenities such as patios, seating areas, water fountains, and wetlands.
STRUCTURAL REQUIREMENTS	14 to 35 pounds per square foot (+/- 70–170 kg/m^2)	59 to 199 pounds per square foot (+/- 290–970 kg/m^2)
GROWING MEDIA	6 inches (15 cm) or less of lightweight growing media	Several feet of soil or specialized growing media
VEGETATION	Relatively low-growing plant communities that can withstand shallow soils, wet and dry conditions, and temperature fluctuations	No restrictions other than those imposed by the conditions of the roof, such as soil depth, exposure, and water availability
IRRIGATION	Dependant on the location and vegetation type	Dependant on the location and vegetation type
MAINTENANCE	Dependent on the vegetation type, typically designed for minimal maintenance	Similar maintenance requirements to a comparable garden at ground level

ADAPTED FROM OBERNDORFER, E. ET AL. 2007.

DESIGN CONSIDERATIONS

Vegetative roofs afford numerous environmental and economic benefits, including urban heat island mitigation, stormwater management, energy conservation, increased roof longevity, sound insulation, the provision of wildlife habitat, and more aesthetically pleasing environments. However, not all vegetative roofs perform equally. Each system should be designed to accomplish specific performance goals and monitored to inform maintenance practices.

INCREASE THE LIFE SPAN OF ROOFS

Green roofs protect the underlying roof membrane from wind, ultraviolet radiation, and other damaging factors. The roof's life span can be increased by two to three times, resulting in significant savings over the life of the building.

BENEFITS TO PHOTOVOLTAIC PANELS

The performance of photovoltaic panels and green roofs can be improved when used in combination. The lower ambient air temperatures provided by vegetative roofs increase the efficiency and yield of photovoltaic panels. The panels, in turn, shade portions of the green roof, improving moisture retention, plant growth, and species diversity (see Figure 4.28).

FRAUENWOHNEN

■ FIGURE 4.28
Native grasses and sedum grow adjacent to solar panels on the FrauenWohnen green roof.

PLANT SELECTION

Similar to any landscape, the most appropriate vegetation is dependent upon the specific conditions of the green roof. Vegetation can be established on the roof using a variety of methods, including seeding, live plants, pregrown vegetative mats, and spontaneous colonization.

Extensive roofs are limited to plant species that can survive in the shallow and often dry growing conditions. Sedums are commonly planted in extensive green roofs; however, other plant species are being successfully used, and the search for suitable vegetation is still underway. Because intensive green roofs contain deeper substrate, they can support a richer plant palette that includes trees and shrubs.

Green roofs are more susceptible to extreme temperature fluctuations than ground-level gardens. As with all gardens, diverse plant palettes provide more resiliency and are recommended due to the susceptibility of monocultures or low-diversity plantings to disease or changes in environmental conditions.

IMPROVE HABITAT

Although all green roofs can create habitat, mimicking local environmental conditions may create more successful and valuable habitat. Construction methods that incorporate local soil, plants, and stone provide a connection to the surrounding ecosystem and can restore habitat for species such as birds, lizards, and insects that were lost during development of the site. Prior to placing native materials on green roofs, tests should be conducted to determine the retention and chemical properties of the soil to ensure the materials can function as required and do not pollute stormwater runoff (Coffman 2009). For more discussion of green roofs and habitat, see Chapter 8, "Sustainable Solutions: Loss of Biodiversity."

IRRIGATION

In order for the full benefits of green roofs to be achieved, the vegetation must remain viable. Depending on the location, soil depth, and vegetation type, green roofs may require supplemental irrigation. Additional water and shade cloth is often required during establishment, particularly when vegetative cuttings or seeds need to be protected and nurtured. A variety of irrigation methods, ranging from drip to capillary systems, can be employed.

In some climates, vegetation can be weaned off additional irrigation after the establishment period. Green roofs that require regular irrigation from potable water should not be considered for a sustainable design in regions with current or impending water shortages. However, the reuse of common wastewater resources, such as rainwater, air-conditioning condensate, or greywater, offer opportunities to support green roofs in more arid climates. Vegetative roofs reusing water in this fashion are also helping to restore the hydrologic cycle of the site and reduce waterwaste. Projects concerned with water use should incorporate a mechanism for tracking water consumption and provide guidance regarding conservation measures in the maintenance plan.

MONITORING AND MAINTENANCE PLAN

Intensive green roofs tend to require more maintenance than extensive systems due to the parklike settings and variety of potential uses. The monitoring and maintenance plan should educate caretakers about the intended design function and components of the green roof system. In addition to standard items, the plan should also include:

- Pollutant-laden maintenance practices that should be avoided, such as the use of certain fertilizers or pesticides that may contaminate roof runoff and impair water quality
- Scheduled inspections of roof drains, gutters, and vegetation-free zones
- Tools or machinery that should be avoided because they may damage the green roof system or waterproof membrane

■ RESOURCES

Cantor, S. 2008. *Green roofs in sustainable landscape design.* New York: W. W. Norton.

Dunnett, N., and N. Kingsbury. 2004. *Planting green roofs and living walls.* Portland, OR: Timber Press.

Snodgrass, E., and L. McIntyre. 2010. *The green roof manual: A professional guide to design, installation and maintenance.* Portland, OR: Timber Press.

Greenroofs.com: www.greenroofs.com

Green Roofs for Healthy Cities: www.greenroofs.org

Livingroofs.org: www.livingroofs.org

■ CASE STUDY

FRAUENWOHNEN (WOMEN'S HOUSING)

PROJECT TYPE: Multifamily residential

LOCATION: Munich, Germany

SIZE: Approximately 0.5 acre (2,000 m)

COMPLETION DATE: 2007

CLIENT: FrauenWohnen EG München

http://www.frauenwohnen-eg.de

THE SITE: The site is part of a mixed-use redevelopment of the former Munich Airport located on the eastern edge of the city.

HIGHLIGHTED SUSTAINABLE PRACTICES:

- Redevelopment of an existing site
- Easy access to public transport
- Rainwater harvesting
- On-site food production
- Zero stormwater runoff
- Extensive stakeholder involvement
- Provides all residents with views of green space
- Site layout encourages social interaction and physical activity
- Flexible layout that allows residents to live and work in the space

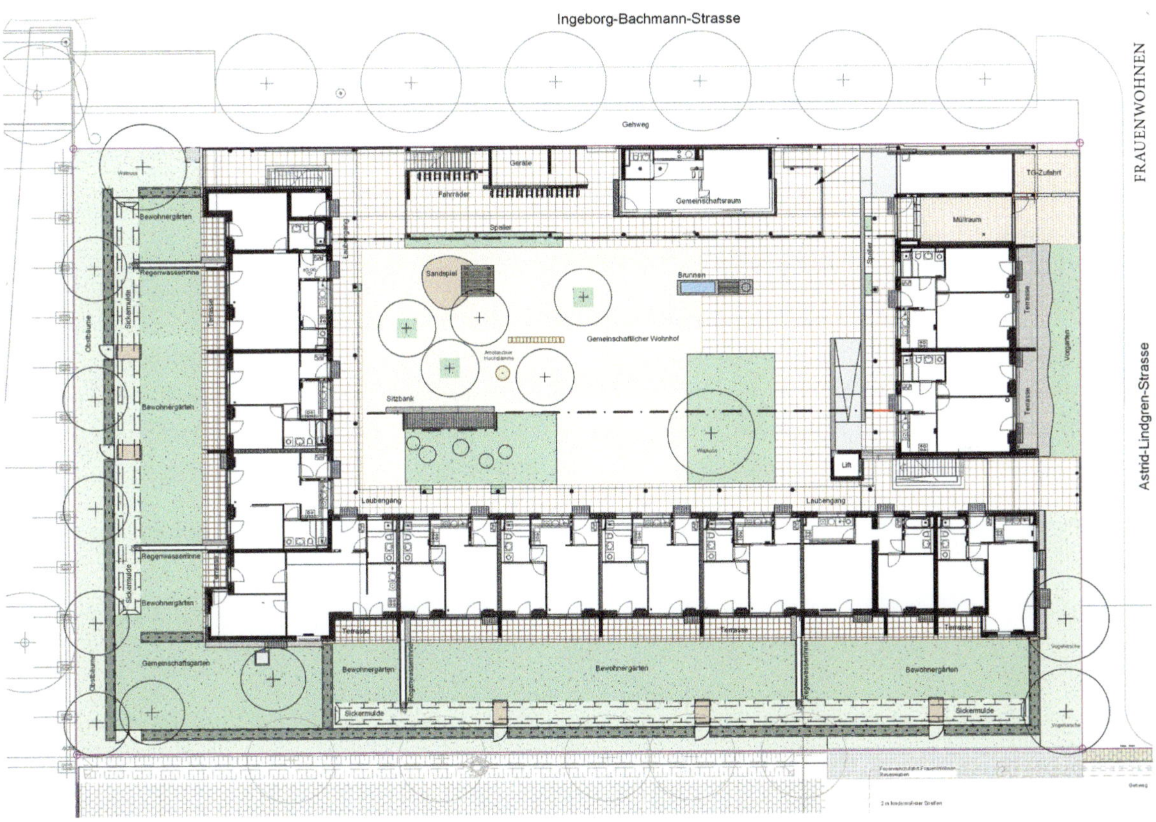

■ **FIGURE 4.29**
FrauenWohnen site plan.

Design Overview

FrauenWohnen was initiated by a women's cooperative that wanted to create affordable housing for women, giving them a chance to live in a loose-knit and mutually supportive community. It consists of forty-nine flats built around a communal courtyard and includes a shared activity room, freelance office space for residents, an apartment for guests, a gym, a workshop, and an administrative building (see Figure 4.29).

continues

FRAUENWOHNEN (WOMEN'S HOUSING) *(CONTINUED)*

The design of the site was developed in consultation with future residents, who gave direct input into all phases of the project. As a result, the residents have an unusually strong connection with both the personal and communal areas of the development. A range of different outdoor spaces were developed within the single small plot to correspond to various needs, including:

- An inner courtyard for communal use (see Figure 4.30)
- A communal front garden
- Private gardens for flats (see Figure 4.31)
- A communal vegetable garden
- Communally owned fruit trees around the outside belt

FRAUENWOHNEN

■ FIGURE 4.30
The inner courtyard offers a generous space for communal activities.

Open access at ground level from the outside to the central courtyard in the south and west encourages neighborhood communication and social interaction. The centerpiece of the courtyard is a water basin, a popular location for residents to gather.

FRAUENWOHNEN

■ FIGURE 4.31
Private gardens behind the ground-floor flats create intimate and personal outdoor spaces for residents and are part of the stormwater management system.

FRAUENWOHNEN (WOMEN'S HOUSING) *(CONTINUED)*

Energy-efficient buildings minimize the need for heating, and solar panels provide an alternative energy source. Green roofs were installed to provide ecological benefits such as on-site stormwater retention and runoff control from roof surface areas, the absorption of air pollution, and the mitigation of the urban heat island.

Water from the roofs is collected in tanks and used to irrigate the gardens and to provide water for the courtyard basin. As a result, the site has zero runoff and reuses 100 percent of its rainwater on-site. Impervious surfaces were kept to a minimum, and the vast majority of the open space is vegetated.

Residents of FraunWohnen decided to opt for a nearly car-free environment. Only twenty-nine out of a potential forty-nine parking spaces were built, of which only eight are used. The residents use a car-sharing scheme and cycle or use public transport.

PROJECT TEAM

- **ARCHITECTS**
  Planungsgemeinschaft Zwischenräume
  Henning, Näbauer, Siedenburg, Meneses
  http://www.zwischenraeume.de
- **LANDSCAPE ARCHITECTS**
  zaharias landschaftsarchitekten + Ulrike Widmer-Thiel
  http://www.zaharias.net
  http://www.ulrikes-gärten.de

Green Walls

Green walls are vegetated wall surfaces. The term is used to describe all forms of vegetated walls, which are typically divided into two major categories: green facades and living walls.

Growing vegetation directly upon walls is an established horticultural practice in many parts of the world. However, the potential damage from root systems attaching to the building surface and the difficulty of routine maintenance activities has limited the practice. To avoid these problems, modern green-wall systems allow vegetation to cover a wall without attaching directly to the building itself.

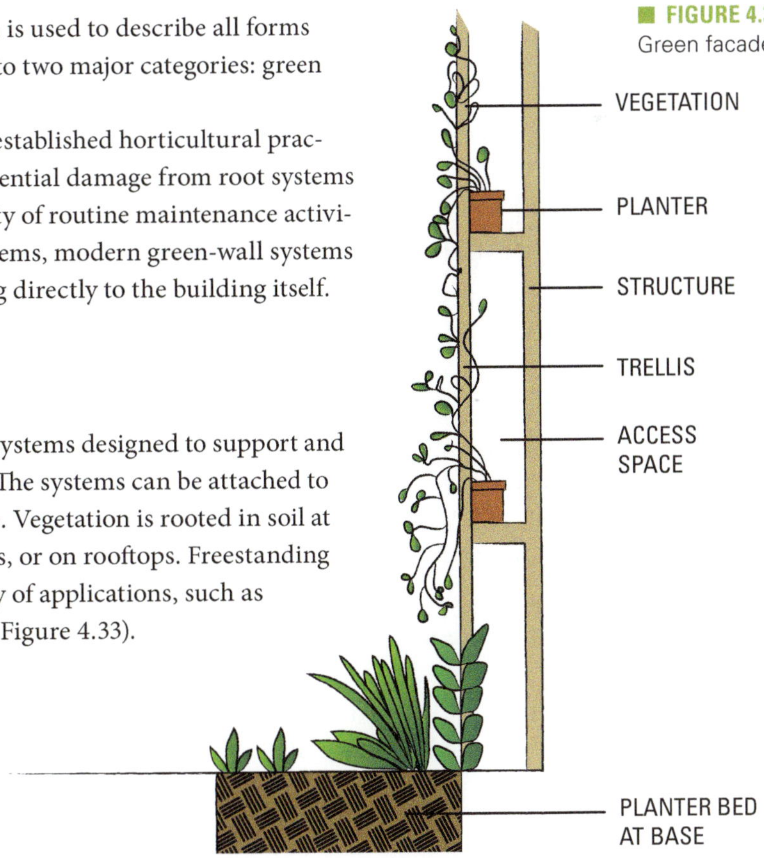

■ **FIGURE 4.32**
Green facade.

GREEN FACADES

Green facades (see Figure 4.32) are vertical wall systems designed to support and be covered by climbing or cascading vegetation. The systems can be attached to existing walls or built as a freestanding structure. Vegetation is rooted in soil at the base of the structure, in intermediate planters, or on rooftops. Freestanding green-wall trellis systems can be used in a variety of applications, such as fences, screens, gazeboes, and exterior walls (see Figure 4.33).

HEATHER VENHAUS

■ FIGURE 4.33 Vines planted in stormwater management terraces grow up a green facade at the Lower Colorado River Authority Redbud Center in Austin, Texas. The vines provide shade on the building's south-facing wall and visually separate the Center from the parking lot.

Options for creating green facades include:

- Wooden or metal trellises
- Cable systems consisting of high-tensile steel cables, wire trellises, anchors, and other supplementary equipment
- Rope made from natural fibers such as hemp (used for short-term projects)
- Freestanding structures such as columns or tree canopy forms made from rigid, lightweight metal panels

DESIGN CONSIDERATIONS

Green facades increase the area available for growing vegetation. In doing so, they can provide a range of economic and environmental benefits, including energy conservation, sound insulation, pollutant removal, wildlife habitat, and enhanced opportunities for mental respite. They can also be used to screen unwanted views and hide unattractive or boring surfaces. Green facades have the added benefit of requiring less soil and space than most trees and can typically be covered in three to five years (Sharp et al. 2008).

The support structure itself can be an attractive landscape amenity whose appearance in various seasons and vegetative growth stages should be considered. Although the weight of the green facade is minimal compared to the weight of the building, not all walls are built to be load-bearing. Professional assistance from the construction and engineering trades should be sought in circumstances where the ability of the wall to support the structure and vegetation is in question. In circumstances where support from the wall is not an option, the full weight of the vegetation can be supported by a rigid system (Dunnett and Kingsbury 2004).

When designing the support structure and selecting the plant palette, it is important to consider the following items:

- Weight of the vegetation and climbing structure
- Additional structural load from climatic variables such as wind, rain, and snow
- Whether the structure will attach to the building envelope or remain freestanding
- Areas of the building facade that should be avoided, such as windows, decorative details, vents, or utility outlets
- Growth characteristic, climbing mechanism, and size of the vegetation. (The maximum height vegetation will climb without requiring additional elevated planters on balconies or ledges to provide water and nutrients is around 78 feet (24 m) (Dunnett and Kingsbury 2004).
- Ability of the vegetation to thrive within the various microclimatic conditions of the wall surface
- Maintenance requirements of both the vegetation and support structure
- Appearance of the vegetation in all seasons

A variety of woody and herbaceous climbing and cascading vegetation can be used in green facades. Select vegetation that can thrive in the conditions of the site and does not require regular irrigation from potable water sources. As with all gardens, opportunities to reuse and recycle on-site water resources such as rainwater, greywater, and stormwater should be explored.

Growing conditions throughout the vertical structure may not be consistent. Changes in the available sunlight, wind, and surface temperatures will impact the vegetation. The design team should investigate any climatic changes that may occur along the vertical structure and select vegetation that can thrive within the harshest conditions.

The monitoring and maintenance plan should include instructions regarding the proper care for both the vegetation and its support structure, and provide guidance for how to determine if the structure may be weakening or causing damage to the building's facade.

■ RESOURCES

Dunnett, N., and N. Kingsbury. 2004. *Planting green roofs and living walls.* Portland, OR: Timber Press.

Green Roofs for Healthy Cities: www.greenroofs.org

Living Walls

Living walls (see Figure 4.34), also known as biowalls or vertical gardens, are typically composed of prevegetated panels, vertical modules, or planted blankets that are attached to a structural wall or frame. Vegetation growing in the wall is rooted in lightweight soil or layers of fibrous materials, such as felt or plastic mesh. Similar to a hydroponic system, a relatively constant supply of water slowly drips through the wall, and a reservoir at the bottom

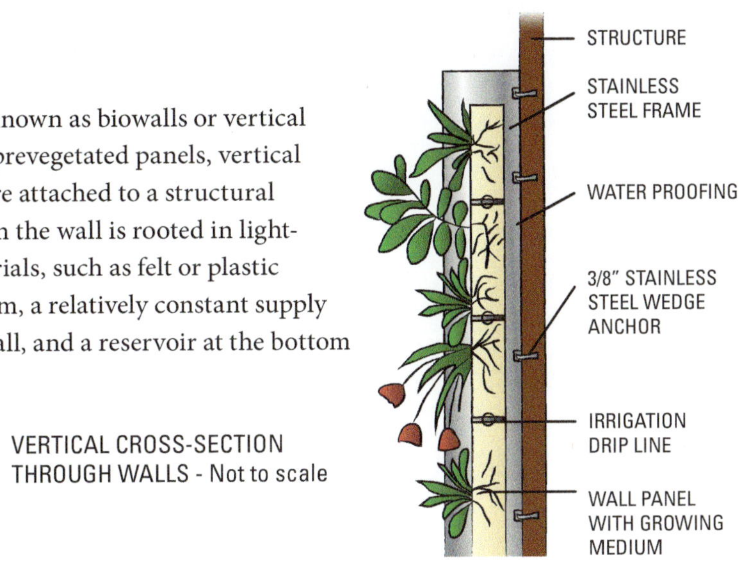

■ **FIGURE 4.34**
Living walls.

of the wall collects and recirculates the water. Nutrients are often added to the irrigation system to help sustain healthy vegetation. Some living walls are designed to be part of an aquatic ecosystem, with a pond located at the base of the wall that serves as both a reservoir and a habitat for plants and fish (Loh 2008).

BILL TIMMERMAN

BILL TIMMERMAN

■ **FIGURE 4.35** Constructed of steel columns, wire mesh, filter fabric, and lightweight soil, the Phoenix Convention Center's living wall is planted with a native seed mix and plant plugs. Designed by Ten Eyck Landscape Architects, the wall receives air-conditioner condensate from the Convention Center. The condensate is pumped outside the building, where it descends through three stainless-steel discs and trickles down rain chains before its journey through the vertical garden. Runoff from the wall is diverted via runnels to the adjacent sunken water-harvesting garden, which also captures rainwater runoff from the adjoining plaza.

■ DESIGN CONSIDERATIONS

Living walls can provide a variety of benefits to buildings and the surrounding microclimate. The vertical gardens can shade and insulate building walls, trap airborne pollutants, reduce noise, and provide an area for urban food production. Through the process of evaporation, the flowing water and vegetation can also cool surrounding air masses, particularly in arid climates with low humidity. Many living wall systems have prevegetated panels, making them capable of providing immediate benefits. Because the vegetated surface is vertical and the height can be adjusted, living walls offer ideal gardening opportunities for children and physically challenged individuals who may have difficulty bending or kneeling.

Living wall systems are resource dependent and many have intense water, energy and nutrient requirements. Projects relying on potable water, nonrenewable energy resources, and intensive maintenance practices can accrue an environmental debt that far outweighs the benefits of the living wall system. In order to determine whether or not a living wall is a sustainable option for a site, the water source, energy use, and long-term maintenance must be considered.

To help ensure success and offer continued learning opportunities, project teams should incorporate mechanisms into the design to monitor performance and resource use (water, energy, fertilizers) of the vertical garden. Without monitoring feedback, maintenance staff can easily overcompensate with potable water or fertilizers, thereby increasing the wall's environmental footprint. The monitoring and maintenance plan should educate caretakers about how to interpret and use the monitoring information to properly maintain and improve the system's function. Guidance should also be provided regarding regular maintenance activities such as plant care and replacement techniques. Living walls are a fairly new technology, and research is still being conducted to determine the longevity of such systems.

■ RESOURCES

Dunnett, N., and N. Kingsbury. 2004. *Planting green roofs and living walls.* Portland, OR: Timber Press.

greenscreen, resources + downloads. "Introduction to Green Walls Technology, Benefits and Design, September 2008." http://www.greenscreen.com/home.html

Loh, S. 2008. Living walls: A way to green the built environment. *Environment Design Guide Technology* 26: 1–8.

Minimize Maintenance Practices That Release Air Pollutants

Over the life of a project, site maintenance can be a substantial and continuous source of air pollution. The general upkeep of most developed landscapes depends upon the regular use of lawn and garden equipment that emit considerable amounts of carbon monoxide, volatile organic compounds, and nitrogen oxides. Zero- or low-emission maintenance practices such as manual tools and propane- or electric-powered mowers can be used as alternatives to gas-or diesel-powered equipment; however, they are not always feasible or without environmental and human health costs. In addition, high-embodied energy maintenance practices, such as the application of potable water and fertilizers, contribute to air pollution, as does the transportation of equipment to and from the site.

Taking all of these factors into consideration, it quickly becomes apparent that an important and necessary approach for reducing the long-term generation of air pollution is to strategically minimize the maintenance of the site. This is not to say that the site should be neglected; rather, the layout, materials, and plant palette should be carefully chosen to reduce maintenance and the required use of land-care practices that release harmful air pollutants. Project teams should explore the maintenance requirements of potential design solutions throughout the development process to find the design options that best support the intended use and ecological function of the site with minimal maintenance.

Design strategies to reduce site maintenance and the release of harmful air pollutants include the following:

- Invite a land-care professional to join the project team and participate in the design process. Weigh design decisions against the long-term environmental and economic maintenance costs and the ability or willingness of the client to maintain the site sustainably.
- Incorporate existing vegetation into the site design to minimize energy use and maintenance during the installation and establishment of new vegetation.
- Select vegetation that can flourish within the built conditions of the site and, once established, requires minimal maintenance.
- Specify durable materials and products that are appropriate for the intended use and will last the life of the project with minimal maintenance. Minimize the release of volatile organic compounds and other air pollutants by avoiding materials that require sealing or coating every few years.
- Limit the use of potable water to applications that require water suitable for drinking. Use alternative water sources, such as stormwater, rainwater, greywater, or air-conditioning condensate, for all other purposes.
- Avoid the use of synthetic fertilizers and pesticides, which often have high embodied energy.
- Infiltrate and reuse stormwater on-site to reduce the treatment load and energy use of wastewater facilities.
- Develop a monitoring and maintenance plan that conveys the activities, maintenance schedule, and equipment required to minimize air pollution and successfully sustain the natural and built components of the site.

DEVELOP A MONITORING AND MAINTENANCE PLAN

The responsibility of a project team for reducing a site's operating energy does not end with its design. All sites require some level of maintenance, and the neglect of a site can lead to an unsustainable cycle of remove, replace, and rebuild that increases environmental and economic costs. Do not assume that future site owners and land-care professionals will automatically maintain the site as envisioned by the project team. To inform and guide the sustainable management of the site, provide a monitoring and maintenance plan to the client. The plan should acknowledge that landscapes are living systems that will change over time and convey the monitoring activities that will inform maintenance practices and the schedule and equipment required to minimize air pollution and successfully sustain the natural and built components of the site. Further discussion of site maintenance plans can be found in Chapter 2, "The Sustainable Site Design Process."

WATER AND ENERGY USE

The discussion of sustainable sites and water typically centers around two issues: dwindling water supplies and the importance of conservation, and on-site stormwater management. Water and its relationship to energy use are often overlooked. However, the collection, distribution, and treatment of potable and wastewater require substantial energy and financial resources (NRDC 2009). In the United States alone, 52,620,000 metric tons of carbon dioxide—the equivalent to the amount generated by ten million cars—is released each year by water treatment facilities (U.S. EPA 2010a). Because the vast majority of water used in a landscape does not need to meet drinking water–quality standards, it is wasteful and unnecessary to use resource-intensive potable water for most landscape purposes. In addition, when stormwater is transported to wastewater treatment plants instead of managed on-site, a perfectly suitable source of water for the landscape is lost, and the treatment load of the wastewater plant increased. Projects that use alternative on-site water resources, such as rainwater, greywater, and air-conditioner condensate, not only reduce air and water pollution but make efficient use of valuable water resources. For a more detailed discussion of the reuse of alternative water sources and strategies to minimize stormwater runoff, see Chapter 5, "Sustainable Solutions: Urban Flooding and Water Pollution," and Chapter 6, "Sustainable Solutions: Water Shortages."

INCORPORATE EXISTING VEGETATION INTO THE SITE DESIGN

Sustainable sites can reduce environmental damage and resource use by incorporating existing on-site native and other site-appropriate vegetation into the design. Replanting a site once construction is complete may not be adequate compensation for the vegetation that is removed. To understand the full costs, look beyond the initial construction efforts and consider other factors, including the environmental, economic, and human health benefits lost due to the temporary or permanent removal of the vegetation; the energy consumed and environmental impacts that occur from the removal, disposal, or reuse of the vegetation; and the energy and environmental impacts required to grow, transport, and establish new plants on-site.

■ **FIGURE 4.36** Nestled beneath a canopy of live oaks (*Quercus virginiana*), Hacienda Ja Ja was carefully designed to limit site disturbance, promote cross-ventilation, and maximize natural daylighting. In order to protect the native trees and minimize disturbance of the root zone, Lake Flato Architects adjusted the building foundation by cantilevering corners and discontinuing the perimeter beam.

FRANK OOMS

Sites can avoid accruing this "environmental debt" by minimizing the area that needs to be revegetated after construction. Integrating existing vegetation into the project can also help establish a unique sense of place and provide immediate habitat, shade, stormwater management, and other benefits. Protecting on-site vegetation may require more effort during the design and construction process; however, the long-term environmental and economic benefits far outweigh the initial conservation costs.

Not all vegetation should be preserved, however. Invasive species, plants that are unhealthy, and vegetation that is not well suited to the conditions of the site can be more of a liability than a benefit to the project and surrounding area. In such a circumstance, replacing the inappropriate or diseased vegetation with healthy species better suited to the site is a more sustainable option.

Successfully incorporating vegetation into the site design requires the input and interaction of an integrated design team. Expertise is needed from design, construction, and maintenance professionals to synthesize site features and ensure the vegetation's protection and long-term health. Plants that are preserved but not adequately protected during construction are more prone to stress, disease, and premature death. As a reference, the Sustainable Sites Initiative recommends protecting vegetation from construction activities with a fence or other physical barrier that cannot be easily moved, and providing the following vegetation and soil protection zones (see Figure 4.37):

- **TREES:** Protect a circular area around each tree that extends out from the tree trunk a distance of 2 feet per inch (0.24 m per cm) of the trunk diameter at breast height (dbh) or the full lateral extent of tree root systems as determined by ground-penetrating radar.
- **SHRUBS:** Protect a circular area around each shrub that extends out from the stem to twice the radius of the shrub.
- **HERBACEOUS VEGETATION:** Protect the entire diameter of the plant.

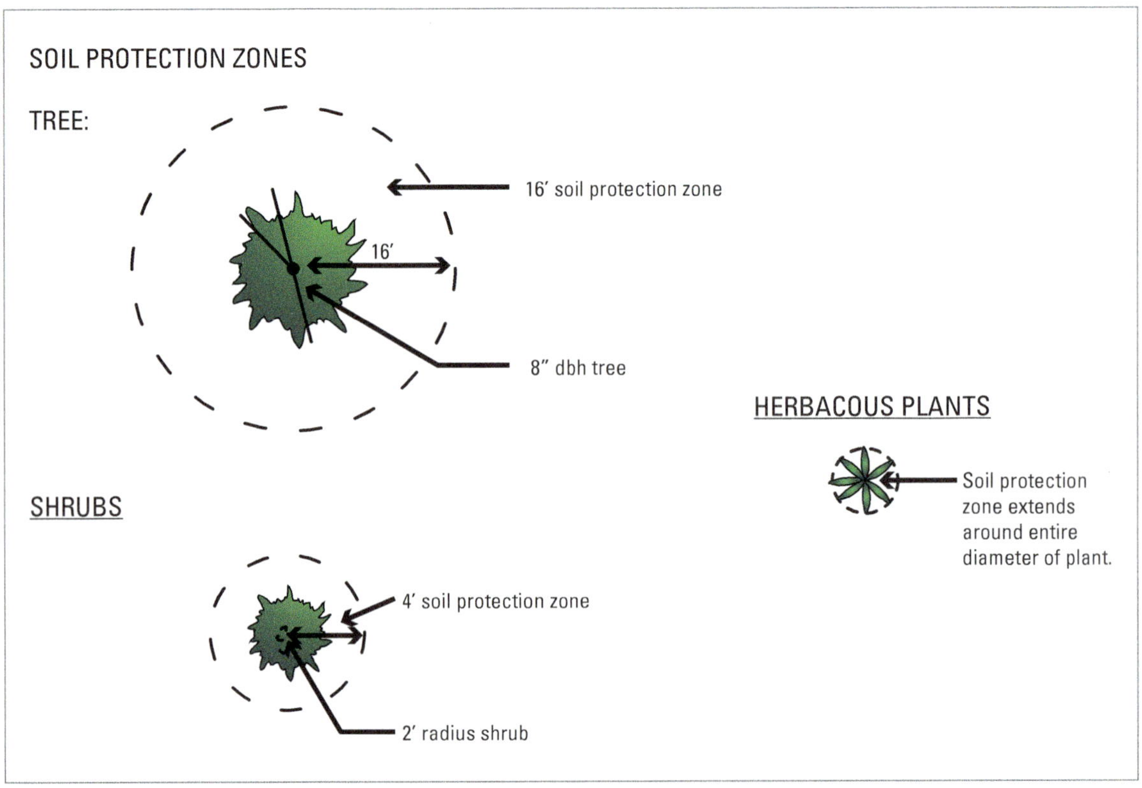

■ **FIGURE 4.37** Recommended soil protection zones.

SELECT VEGETATION WITH LOW MAINTENANCE REQUIREMENTS

Vegetation is a versatile and key component of a sustainable site essential to carbon sequestration, clean air, climate regulation, and numerous other ecosystem processes. The air quality benefits provided by the plants growing on-site can be outweighed by the embodied energy and greenhouse gas emissions from such maintenance activities as potable water irrigation, fertilizer and pesticide application, and the use of fossil fuel–powered machinery (Small and Czimczik 2010). To avoid this counterproductive and unsustainable outcome, plants selected for the site should be able to flourish within the built conditions and, once established, require minimal maintenance. The following strategies will help guide the plant selection process and reduce the long-term maintenance and energy use of the site:

- Select vegetation that is well suited to the soil conditions and can flourish without additional fertilizers or soil amendments.
- Plant perennial and long-lived vegetation.
- Choose slow-growing species that require little pruning. Maintain natural growth patterns, as hedging, topping, and shearing can encourage excessive growth and increase maintenance requirements.
- Select vegetation with minimal irrigation requirements.
- Plant hardy species that are not prone to disease or insect damage.
- Choose vegetation that is fruitless or does not produce large quantities of messy seeds or other materials that will need to be gathered and removed. Such species are often important to wildlife habitat and can be still incorporated into sites where tidiness is a concern by planting in areas where the spread of seeds or other materials does not interfere with major site uses.
- Reduce or eliminate maintenance-intensive lawns. Most turf grass has considerably higher energy requirements than ground covers, shrubs, or trees (Pitt 1984) due to the fertilization, irrigation, herbicide application, and mowing requirements.

A project may nonetheless require the selection of maintenance-intensive vegetation in certain areas. For example, sites that are intended for children or pets may need patches of turf grass for play; however, other surfaces or ground covers may also be suitable. Defining how and when an area will be used can help designers limit maintenance-intensive vegetation to locations where it is absolutely necessary. Design cues that demarcate the transition between areas of high and low maintenance, such as pathways or small walls, will help site users understand the intentional change and guide land-care professionals to the proper maintenance requirements of each zone (see Figure 4.38).

HEATHER VENHAUS

FIGURE 4.38
Trails at the Mueller mixed-use development in Austin, Texas, demarcate the transition between formal perennial gardens and restored native prairie.

Sequester Atmospheric Carbon

The rapid and unprecedented rate at which the global climate is changing has become an international issue of great concern. Atmospheric changes due to substantial increases in greenhouse gases are causing the planet to warm and have the potential to cause impacts such as rising sea levels, freshwater shortages, and declining crop yields.

Sites can mitigate air pollution and the release of carbon dioxide and other greenhouse gases by reducing their embodied and operating energy; they can also sequester atmospheric carbon in vegetation and soil. Carbon sequestration is an important long-term land management strategy and a key component of climate regulation.

Through photosynthesis, vegetation temporarily removes carbon dioxide from the atmosphere and stores carbon in live plant tissues such as leaves, branches, and roots. The carbon is re-released into the atmosphere when the plant tissue decomposes. The length of time aboveground carbon remains sequestered is directly related to the life span of the vegetation and the decomposition rate of the plant tissue. Roots, earthworms, small mammals, and other soil biota incorporate vegetative material and other organic matter into the ground, where it becomes part of the soil structure and provides a variety of benefits, including plant nutrients and increased water-holding capacity.

Once it is bound up physically or chemically within the soil, organic carbon is generally much more stable and long-lived than aboveground carbon because it becomes inaccessible to microbial decomposers. When left physically undisturbed, soil carbon may remain sequestered for many hundreds or even thousands of years. Soil disturbance increases the decomposition rate of organic matter, which, in turn, speeds the release of carbon. Once lost, soil carbon pools can take decades or centuries to rebuild (Jo and McPherson 1995; Potter et al. 1999). Maintaining or increasing vegetative biomass on-site can slowly rebuild carbon pools; however, thoughtful plant selection and maintenance is required to ensure the site remains a net carbon sink.

Strategies for protecting carbon pools and improving carbon sequestration include the following:

- Minimize soil disturbance and erosion.
- Plant long-lived woody vegetation.
- Maintain or increase the vegetative biomass. Select vegetation that is well suited to the conditions of the site and, once established, can be sustained without fertilizers, pesticides, and potable water sources that have high embodied energy.
- Increase the site's vegetative diversity.
- Maintain a site with zero- or low-emission maintenance practices.

■ CASE STUDY

REDSTONE CANYON GARDEN

PROJECT TYPE: Single-family residential

LOCATION: Redstone Canyon near Masonville, Colorado

SIZE: Approximately ⅔ acre (0.8 hectare)

COMPLETION DATE: 2002

CLIENT: Lauren Springer Ogden

HIGHLIGHTED SUSTAINABLE PRACTICES:

- Low-maintenance site with zero emissions
- All garden debris composted and recycled on-site
- Use of local or indigenous materials
- Drought-adapted vegetation is well suited to the growing conditions of the site and does not require fertilizers or soil amendments.

■ **FIGURE 4.39**
Retaining wall and garden path. Low-stature vegetation was selected to better withstand the windy conditions of the site and to limit garden debris and overall maintenance.

THE SITE: The rural site is located on thin and rocky soils in the foothills of Colorado's Rocky Mountains at an altitude of 6,500 feet (1,981m). Chaparral, shortgrass prairie, and coniferous montane plant communities merge in the surrounding area, and large glacial boulders abound. Site challenges included a dry climate with an average annual precipitation of 14 inches (356 mm), mule deer, a threat of wildfire, and a limited well water supply that is not always available.

Design Overview

The high-altitude residential site includes a series of beautiful and multifunctional outdoor living areas that blend seamlessly into the surrounding natural environment. The design is plant-driven and relies on vegetation to vividly express form, texture, and year-round appeal. Inspiration for the expressive gardens came from neighboring plant communities and native rock found on-site.

Outdoor spaces were defined by both practical design requirements and a desire to create a regionally vernacular garden. Plants were chosen according to the unique character of each garden and include native vegetation or other low-water-usage plants that visually blend with the surrounding landscape and provide seasonal color and wildlife value. An emphasis was placed on low-stature vegetation that can withstand the site's windy conditions and also limit garden debris and overall maintenance (see Figure 4.39). All plants thrive in the native soil without imported amendments or fertilizers.

continues

REDSTONE CANYON GARDEN *(CONTINUED)*

Soil is protected from wind and water erosion with either plants or wind-resistant pea gravel mulch. The stone and gravel was either harvested on-site or purchased from a source less than 10 miles from the site (see Figure 4.41). Over 50 percent of the plants were propagated on-site from seed. The vast majority of the remaining vegetation was purchased from nurseries within 100 miles of the site (see Figure 4.40).

■ **FIGURE 4.40** Colors and textures provide year-round appeal. Over 50 percent of the plants were propagated on-site from seed. The vast majority of the remaining vegetation was purchased from nurseries within 100 miles of the site.

■ **FIGURE 4.41** Soil is protected from wind and water erosion with plants and wind-resistant pea gravel mulch. The stone and gravel was either harvested on-site or purchased from a source less than 10 miles from the site.

REDSTONE CANYON GARDEN *(CONTINUED)*

Swales direct rainwater runoff from the house into the garden, and periodic irrigation is applied with soaker hoses. The zero-emissions landscape is maintained by the homeowners without the use of power tools or synthetic pesticides. Guinea fowl are used for grasshopper control, and electric fencing for deer; cats and dogs help to deter rodents. Weeds are controlled primarily through hand digging, vinegar spray, or a flame torch, depending on their location and the vegetation surrounding the area.

PROJECT TEAM

- **LANDSCAPE DESIGNER**
  Lauren Springer Ogden, principal, Plant Driven Design
  www.plantdrivendesign.com
- **PLANT INSTALLATION**
  Lauren Springer Ogden
- **MASONRY INSTALLATION**
  Jesse Young and Ivan Andrade

■ RESOURCES

USDA Forest Service, Climate Change Resource Center. Tree Carbon Calculator. http://www.fs.fed.us/

REFERENCES

American Lung Association. 2010. *State of the air 2010.* Washington, DC: American Lung Association.

Athena Institute, EcoCalculator, http://www.athenasmi.org/tools/ecoCalculator/, accessed 10.11.11.

Baird, C. 1999. *Environmental chemistry.* New York: W. H. Freeman.

Bassuk, N., J. Grabosky, and P. Trowbridge. 2005. *Using CU-structural soil in the urban environment.* Ithaca, NY: Cornell University Department of Horticulture.

Brown, R., and T. Gillespie. 1995. *Microclimatic landscape design: Creating thermal comfort and energy efficiency.* Hoboken, NJ: John Wiley & Sons.

Calkins, M. 2009. *Materials for sustainable sites.* Hoboken, NJ: John Wiley & Sons.

_______. 2011. *The sustainable sites handbook: A complete guide to the principles, strategies and best practices for sustainable landscapes.* Hoboken, NJ: John Wiley & Sons.

Celestian, S., and C. Martin. 2004. Rhizosphere: Surface and air temperatures patterns at parking lots in Phoenix, Arizona, U.S. *Journal of Arboriculture* 30 (4): 245–52.

Coffman, R. 2009. Green roof case study. *Landscape Architecture (January 2009): 72-77.*

Cross, R., and R. Spencer. 2009. *Sustainable gardens.* Collingwood, Victoria, Australia: CSIRO.

Dines, N., and K. Brown. 2001. *Landscape architect's portable handbook.* New York, NY: McGraw-Hill.

Dunnett, N., and N. Kingsbury. 2004. *Planting green roofs and living walls.* Portland, OR: Timber Press.

European Environment Agency. 2010. "About environment and health." Retrieved March 20, 2010, from http://www.eea.europa.eu/themes/human/about-environment-and-health.

Intergovernmental Panel on Climate Change [IPCC]. 2007. *Climate change 2007: Synthesis report summary for policymakers.* Valencia, Spain: Intergovernmental Panel on Climate Change.

Jo, H.-K., and E. G. McPherson. 1995. Carbon storage and flux in urban residential greenspace. *Journal of Environmental Management* 45: 109–33.

Hopper, L. J. 2007. *Landscape architectural graphic standards: Student edition.* Hoboken, NJ: John Wiley & Sons.

Karl, T., J. M. Melillo, T. C. Peterson, and S. J. Hassol. 2009. *Global climate change impacts in the United States.* New York: Cambridge University Press.

Loh, S. 2008. Living walls: A way to green the built environment. *Environment Design Guide Technology* 26.

McPherson, G. 1984. "Solar control planting design." In *Energy-conserving site design.* American Society of Landscape Architects.

Miller, G. T. 1998. *Sustaining the earth: An integrated approach.* Wadsworth.

National Aeronautics and Space Administration [NASA]. 2010. "Earth observatory: Global warming." Retrieved January 5, 2011, from http://earthobservatory.nasa.gov/Features/GlobalWarming/.

National Institute of Standards and Technology, BEES, http://www.nist.gov/el/economics/BEESSoftware.cfm, accessed 10.11.11.

Natural Resource Defense Council [NRDC]. 2009. *Water efficiency saves energy: Reducing global warming pollution through water use strategies.* New York: Natural Resource Defense Council.

Ngo, C., and J. Natowitz. 2009. *Our energy future: Resources, alternatives and the environment.* Hoboken, NJ: John Wiley & Sons.

Nowak, D. J., D. E. Crane, and J. C. Stevens. 2006. Air pollution removal by urban trees and shrubs in the United States. *Urban Forestry and Urban Greening* 4: 115–23.

National Renewable Energy Laboratory [NREL]. 1995. *Landscaping for energy efficiency.* Washington, DC: U.S. Department of Energy.

Oberndorfer, E., J. Lundholm, B. Bass, R. R. Coffman, H. Doshi, N. Dunnett, S. Gaffin, M. Köhler, K. K. Y. Liu, and B. Rowe. 2007. Green Roofs as Urban Ecosystems: Ecological Structures, Functions, and Services. *BioScience* 57(10): 823–33.

Pitt, D. 1984. "Conservation of embodied energy through landscape design." In G. McPherson, *Energy-conserving site design.* American Society of Landscape Architects.

Pomerantz, M., H. Pon, H. Akbari, and S.-C. Chang. 2000. *The effect of pavement temperatures on air temperatures in large cities.* Berkeley, CA: Lawrence Berkeley National Laboratory.

Potter, K. N., H. A. Torbert, H. B. Johnson, and C. R. Tischler. 1999. Carbon storage after long-term grass establishment on degraded soils. *Soil Science* 164 (10): 718–25.

Reed, S. (2010). *Energy-wise landscape design: A new approach for your home and garden.* New Society.

Sharp, R., J. Sable, et al. 2008. *Introduction to green walls: Technology, benefits, and design.* Green Roofs for Healthy Cities.

Thompson, J. W., and K. Sorvig. 2000. *Sustainable landscape construction: A guide to green building outdoors.* Washington, DC: Island Press.

Townsend-Small, A., and C. I. Czimczik. 2010a. Carbon sequestration and greenhouse gas emissions in urban turf. *Environmental Sciences* 37 (2): 1–5.

———. 2010b. Carbon sequestration and greenhouse gas emissions in urban turf. *Geophysical Research Letters* 37.

U.S. Department of Energy. 2010. "Energy savers: Landscape shading." Retrieved October 2, 2011, from http://www.energysavers.gov/your_home/landscaping/index.cfm/mytopic=11940.

U.S. Energy Information Administration [EIA]. 2010. "General energy." Retrieved December 19, 2010, from http://www.eia.doe.gov/ask/generalenergy_faqs.asp#energy_use_buildings.

U.S. Environmental Protection Agency [EPA]. 2003. Reducing air pollution for nonroad engines. *EPA* 1–12.

———. 2007. Inventory of U.S. Greenhouse Gas Emissions and Sinks, 1990–2005. April 2007. Report EPA 430-R-07-002.

———. 2008. "Effects of acid rain: Surface waters and aquatic animals." Retrieved January 20, 2011, from http://www.epa.gov/acidrain/effects/surface_water.html#a2.

———. 2010a. *National water program strategy: Response to climate change.* Washington, DC: U.S. EPA Office of Water.

———. 2010b. "What is acid rain?" Retrieved February 10, 2011, from http://www.epa.gov/acidrain/what/index.html.

RONALD HUDSON/BIGSTOCK.COM

■ **FIGURE 5.1**
Flooding in North Yorkshire, UK.

CHAPTER 5

Sustainable Solutions: Urban Flooding and Water Pollution

HUMANITY HAS A LONG, complex—and interdependent—relationship with water. Human civilization itself was born in a place we now call the Fertile Crescent, an area swaddled by the Tigris, the Euphrates, and the Nile rivers. As civilized society spread throughout the globe, cities sprouted along the banks of waterways: Shanghai at the mouth of the Yangtze, Paris on the river Seine, Kanpur on the Ganges, Rome on the Tiber, and New Orleans at the mouth of the mighty Mississippi. Water bodies attract development because of the many advantages they offer in transportation, commerce, energy production, food, and recreation. With these benefits, however, come disadvantages—primarily flooding and water pollution.

Flooding is a natural process. It is necessary for maintaining the function and biodiversity of many aquatic and terrestrial ecosystems. Floodwaters create critically important habitat, return nutrients to the land, recharge groundwater supplies, and replenish topsoil. Flooding becomes problematic, however, when property damage occurs and people's homes and businesses are affected.

Land-use changes associated with urban development often exacerbate flood conditions and are a major source of water pollution. Impervious surfaces, which cover significant portions of our built environment, increase both stormwater volume and runoff rate, which adds to the pollution of the vital waterways on which we depend.

Clean water legislation has been effective at slowing the degradation of, and in some cases even improving, water quality; however, there is still much work to be done. In the United States, half of the rivers and streams—roughly 463,000 river and stream miles (745,126 km); 66 percent of the lakes, reservoirs, and ponds—equaling approximately 11 million acres (4.5 hec); and 100 percent of the Great

Lakes open waters—an area of 56,709 square miles (146, 875 km^2)—have been classified as "impaired" (U.S. EPA 2011), meaning they have been polluted to the point that they no longer meet water quality standards for their designated use. And most state agencies are able to sample or monitor only a small percentage of their waters. A more comprehensive look at all aquatic ecosystems in the United States would likely uncover higher percentages of impairment.

The good news is that sustainable site development can reduce urban flooding and improve water quality by restoring the ecosystem processes that capture and cleanse water. In this chapter, the relationship between urban flooding, water pollution, and site development is explored. Pollutant sources and their impacts on human health and the environment are discussed, along with strategies to restore floodplain function, reduce stormwater runoff, and improve water quality.

■ POINT SOURCE AND NONPOINT SOURCE WATER POLLUTION

Water pollution can be classified as either point source or nonpoint source. Pollution discharged into a body of water from a discernible, confined, and specific location, such as a pipe, ditch, or sewer, is defined as point source water pollution. Because it originates from a discrete location, it is typically easier to trace, monitor, and control. Point source pollution is typically associated with industrial water discharges and sewage treatment plants.

Nonpoint source water pollution is dispersed and is not attributable to a single point of discharge. Polluted runoff from landscapes, roads, and parking lots are common examples. Because it originates from many different sources across a broad geographic area, nonpoint source pollution is more difficult to monitor and control.

In the United States, initial efforts to improve water quality focused on point source pollution. More recently, attention has shifted to nonpoint sources due to an improved understanding of pollutants and the increasing volume of stormwater runoff from urban environments.

Flooding and Water Pollution

The three leading factors related to site development that contribute to urban flooding and water pollution are (1) the development or alteration of floodplains and the subsequent loss of healthy floodplain functions, (2) impervious surfaces and the resultant increases in stormwater runoff, and (3) combined sewer overflows.

The Cause: Floodplain Development

Floodplains are the lowlands and relatively flat areas adjoining inland and coastal waters that are subject to flooding. They are an extension of the water system; their function is to convey high water volumes downstream or temporarily store water until flooding subsides.

Fully functioning floodplains provide a variety of ecosystem services including:

- Floodwater storage and peak flow moderation resulting in the reduced severity of floods
- Filtration and removal of water pollutants
- Channel stability and erosion control
- Wildlife habitat
- Groundwater recharge
- Stream baseflow
- Beauty
- Recreational opportunities

Development or alteration of—or encroachment on—floodplains can disrupt or greatly reduce their ability to provide these valuable environmental and economic benefits.

Floodplains are classified according to the likelihood of flooding in a given year. For example, the one-hundred-year floodplain demarcates the elevation that has a 1 percent chance of being flooded each year. The term can be misleading and is commonly mistaken for a flood that will occur once every hundred years. In reality, the flooding can occur multiple times within a relatively short time period. Because flood events are not always consistent in their timing or severity, risks are often ignored or downplayed until flooding occurs and it is too late to prevent poor development choices.

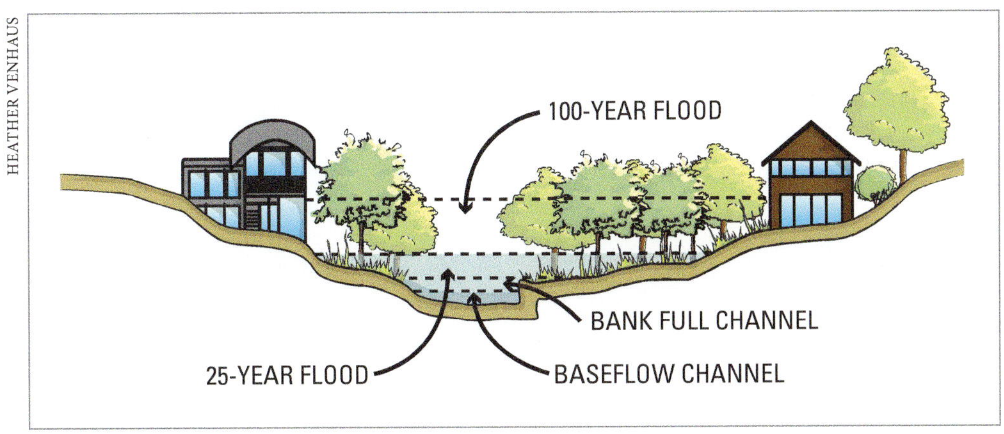

■ **FIGURE 5.2**
River floodplain cross-section.

Flooding: How It Affects Our Lives

The development and alteration of floodplains has resulted in an increase in flooding, danger to humanity, and environmental degradation. Flood damages in the United States average over $3 billion each year (NOAA 2010) and account for about 39 percent of all deaths from natural disasters—more than any other type (Miller 1998).

The Cause: Impervious Surfaces and Stormwater Runoff

Stormwater runoff is rain or snow melt that flows off the site instead of soaking into the ground. Impervious surfaces are the primary cause of stormwater runoff in urban environments. Conventional roads, roofs, parking lots, and other impervious surfaces decrease the area of vegetated land available to infiltrate rainwater, which in turn increases the stormwater volume and runoff rate of the site (see Figure 5.3). This disruption of the hydrologic cycle degrades the quality and reduces the quantity of water resources by limiting groundwater recharge and transporting pollutants from urban land to nearby waterways.

■ FIGURE 5.3
Impervious cover and stormwater runoff.

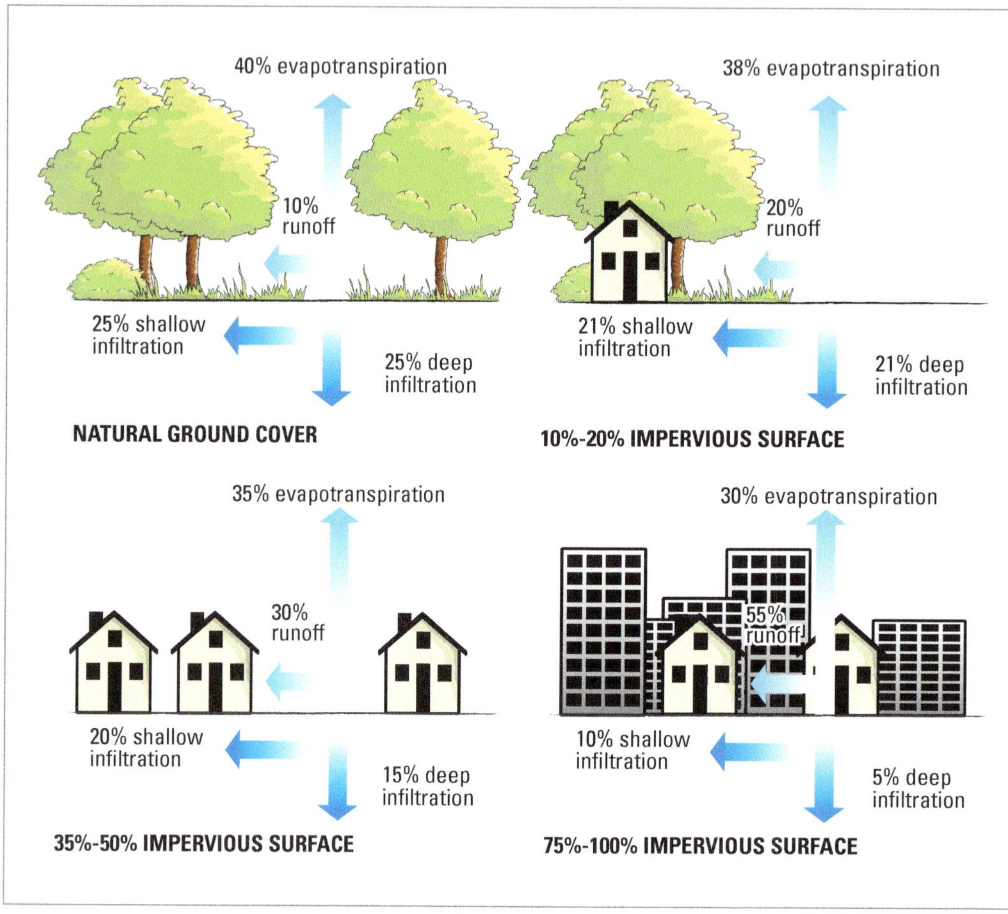

FEDERAL INTERAGENCY STREAM RESTORATION WORKING GROUP

Common stormwater pollutant sources include:

- Fertilizers, herbicides, and insecticides
- Animal waste
- Road salt
- Coal tar-based sealants used on paved roads
- Vehicle fluids, exhaust, brake linings, and tire and engine wear
- Sediment from improperly managed landscapes
- Roofing materials
- Debris

Stormwater Runoff: How It Affect Our Lives

Stormwater runoff from developed land is the leading cause of water pollution in urban areas (Loizeaux-Bennet 1999). As the stormwater moves across urban surfaces, it increases in temperature and accumulates sediment, nutrients, pathogens, heavy metals, and other pollutants that adversely affect water quality and degrade downstream aquatic habitats. In addition, stormwater runoff increases the volume and rate of water entering sewer systems and stream channels, increasing the likelihood of flooding and the spread of pollutants.

Conventional storm sewer systems typically utilize pipes, culverts, or channels to quickly and discretely discharge stormwater runoff to receiving water bodies such as lakes, streams, and wetlands. The runoff is not treated and pollutes water bodies used for drinking and recreation, as well as wildlife habitat (see Table 5.1).

■ **TABLE 5.1** Common Stormwater Pollutants

STORMWATER POLLUTANT	SOURCE	IMPACT
SEDIMENT	Disturbed or bare soils	Sediment reduces water quality and degrades aquatic habitat. Nutrients, metals, and other pollutants can attach to, and are transported by, sediment.
NUTRIENTS	Animal waste, failing septic systems, and fertilizers	Elevated nutrient loads reduce water quality and degrade aquatic habitat by stimulating algal blooms, lowering dissolved oxygen levels in water bodies, and reducing water clarity. Nutrients also increase water treatment costs.
BACTERIA	Animal waste, combined sewer overflows, failing septic systems	Harmful to the health of humans and wildlife.
TEMPERATURE	Replacing vegetation with dark and impervious surfaces such as roads, driveways, and roofs	Significantly impacts populations of fish, particularly cold-water species of salmon and trout, by lowering dissolved oxygen levels in water bodies.
METALS	Pesticides, herbicides, roofing materials, tires, brake dust, automobile engine wear, fuel, asphalt paving	Harmful to the health of humans and wildlife, even at low levels.
CHLORIDE	Road deicing salts	Contaminate soils and water, and harm vegetation and aquatic wildlife

The Cause: Combined Sewer Overflows

Combined sewer systems (CSS) collect stormwater and sewer and industrial waste into the same pipeline and transport the wastewater to a treatment plant (see Figure 5.4). During intense rain events, stormwater runoff can overwhelm the capacity of such systems. When this occurs, the CSS overflows and discharges excess wastewater into nearby streams, lakes, or other water bodies (U.S. EPA 2001c).

Combined sewer systems serve an estimated forty-three million Americans and are primarily located in older cities in the Northeast and the Great Lakes region of the United States; a small percentage are also along the West Coast (U.S. EPA 2001c). Combined systems are also common throughout Europe and make up a majority of the sewerage systems in the UK (DEFRA 2010).

Replacing combined sewer systems with separate storm and sanitary sewers is an expensive and lengthy process. Many municipalities are successfully working with property owners to reduce stormwater runoff, and in doing so have avoided increases in infrastructure costs.

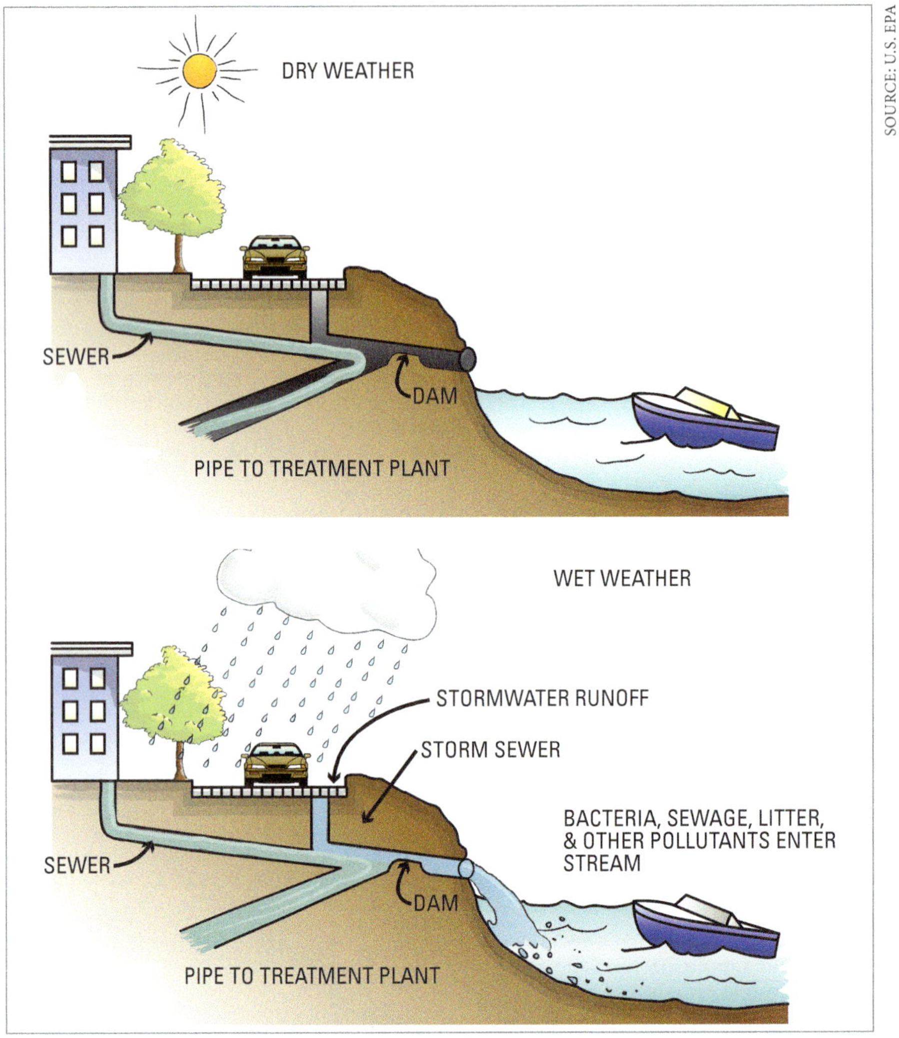

■ FIGURE 5.4 Combined sewer system.

Combined Sewer Overflows: How They Affect Our Lives

Combined sewer overflows (CSO) release stormwater pollutants and untreated human and industrial waste, toxic materials, and debris into nearby rivers, lakes, and other water bodies. The untreated wastewater is released in order to prevent backup into streets, homes, and businesses served by the combined sewer system. The overflows can greatly diminish water quality and are a health risk for humans and wildlife. CSOs also limit the aesthetic value and enjoyment of waterways. Common pollutants include:

- Bacteria such as fecal coliform and *E. coli*
- Viruses such as hepatitis and diphtheria
- Parasites such as giardia and cryptosporidium

- Metals such as lead, zinc, cadmium, and chromium
- Oil and grease
- Trash and litter
- Nutrients such as nitrogen and phosphorous (U.S. EPA 2001c)

Sustainable Site Strategies to Mitigate Urban Flooding and Water Pollution

Sustainable sites mitigate urban flooding and water pollution by protecting and restoring the capacity of a site to cleanse and temporarily store water. Design teams can hold water on-site using both built (e.g., porous paving and green roofs) and natural (vegetation and soil) site components that intercept, infiltrate, and evaporate rainwater. Rainwater can also be harvested and stored in cisterns or other containers for reuse on the landscape or in buildings.

Existing sources of stormwater runoff and water pollution, such as parking lots or heavily fertilized lawns, should be identified in the site inventory, and design strategies developed to capture and cleanse the water on-site. Vegetation, soil, and the diverse community of microorganisms that live within the soil can bind and break down many water pollutants. Sites can take advantage of these natural cleansing mechanisms by using design strategies such as biofiltration areas that slow runoff and filter it through vegetation and soil.

At the outset of a project, goals and performance targets should be established for reducing stormwater runoff and avoiding the use of materials or products that pollute water resources. New design opportunities arise when project teams view stormwater not as a waste product, but as a valuable resource that can be utilized to improve the function and beauty of the site.

Site strategies to mitigate urban flooding and water pollution include the following:

- Avoid the development of flood prone areas
- Restore previously developed floodplains
- Reduce impervious surfaces
- Protect and restore soil health
- Increase vegetative cover
- Slow stormwater runoff and improve infiltration

Avoid Development of Flood Prone Areas and Restore Previously Developed Floodplains

The most straightforward strategy for reducing urban flooding and water pollution is to avoid development of flood-prone areas and restore previously developed or degraded floodplains. When selecting a site, priority should be given to landscapes that have already been developed, such as greyfield and brownfield properties. The redevelopment of degraded sites diverts development away from greenfields and provides the opportunity to restore the many benefits provided by healthy, functioning ecosystems.

Floodplain restoration practices vary and are dependent on the floodplain type and condition of the ecosystem. Project teams tackling this issue should include landscape ecologists, hydrologists, environmental engineers, or other professionals who specialize in the restoration of floodplains and can determine the most effective strategies.

When restoring floodplains, the site inventory and analysis often extends to the watershed level to identify conditions that influence the function and overall health of the floodplain. Because the on-site biotic function is heavily influenced by off-site factors, the opportunity to fully restore a floodplain to full functionality is rare. However, all floodplain restoration projects can improve biological function relative to its baseline conditions and help reduce flood concerns. Examples of restoration practices include:

- Removing materials or surfaces that release pollutants into the waterway
- Removing impervious surfaces and restoring soil health
- Reconfiguring channelized or degraded stream banks and shorelines
- Removing invasive species
- Stabilizing slopes, stream banks, and shorelines with vegetation or other soft engineering practices
- Restoring native plant communities appropriate to site conditions

Where development of an intact floodplain cannot be avoided, the project team should begin the design process by investigating the full environmental, economic, and social benefits provided by the floodplain and identify opportunities to maintain these ecosystem services. Options may also exist for off-site mitigation within the watershed to minimize the overall loss of floodplain functions. Mitigation should only be considered after all other efforts to avoid greenfield development and protect existing ecosystem services on-site have been exhausted. See Chapter 8, "Sustainable Solutions: Loss of Biodiversity," for a more detailed discussion of mitigation.

All development is not incompatible with floodplain ecosystems. Land-use practices such as parks or trail systems that protect floodplain functions and maintain or improve water quality are often appropriate land-use options.

Monitoring and management of the floodplain is a key factor to the long-term success of a project. Overuse or misuse can compromise the ecological integrity of a site. Plans should be put in place to monitor the impacts of floodplain use and relocate damaging activities as needed so that the site can naturally rehabilitate or be actively restored.

Reduce Impervious Surfaces

A significant portion of our urban environments are covered by impervious surfaces, which are the leading cause of stormwater runoff and a major contributor to water pollution (see Figure 5.5). Reducing the volume and velocity of runoff decreases the likelihood of flooding and the pollution of groundwater and other receiving water bodies. Airborne pollutants, automobile fluids, and other contaminants accumulate on impervious surfaces. During the first portion of a rain event, the bulk of the pollutant load is carried away by the "first flush" of stormwater. Containing and infiltrating the first flush on-site allows pollutants to be treated at the source, helping to prevent the spread of contaminants.

In addition to impervious cover, soils and vegetation type can also impact the volume of runoff from a site. Vegetation with fine, dense surface roots such as those commonly found in turfgrass lawns can dramatically limit the infiltration capacity of the soil (Urban 2008). Runoff volume can also increase when soils become overly compacted and lose their ability to absorb water. Common pollutants associated with runoff from vegetated surfaces include phosphorus, fecal coliform, and sediment (Bannerman et al. 1993).

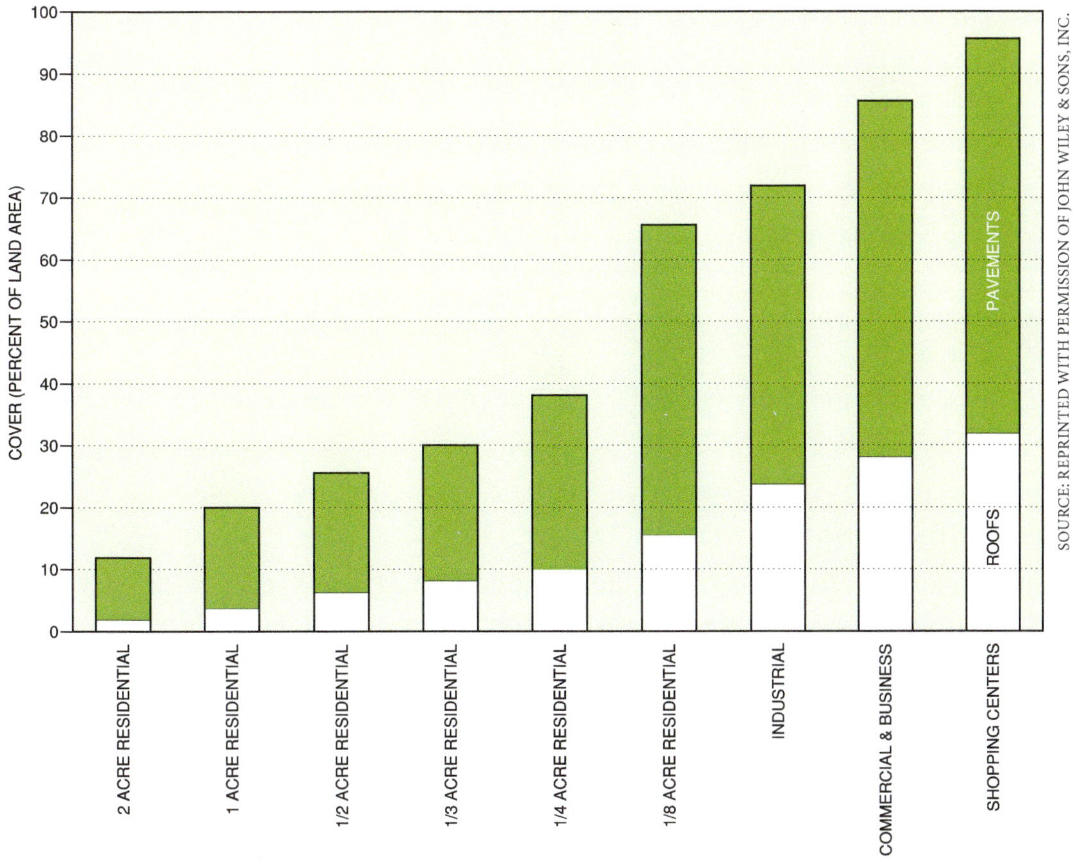

■ **FIGURE 5.5** Impervious coverage in typical land uses.

When runoff volumes are reduced, less infrastructure is required to transport and treat stormwater, resulting in cost savings for the developer and residents. A good example of this is the Somerset subdivision in Prince George's County, Maryland, which used a combination of low-impact development (LID) strategies, including bioswales and rain gardens, to mange stormwater. The portion of the subdivision that used the LID approach reduced stormwater volume by approximately 20 percent, saving $785,382, a 32 percent savings over conventional development costs (U.S. EPA 2007).

Sustainable site strategies to reduce impervious surfaces include:

- Pave less
- Permeable paving
- Structural soils
- Green roofs

Pave Less

Overpaving is a common characteristic of urban environments. Many cities are dominated by automobilecentric designs, with wide roads and expansive parking lots that result in safety and health hazards for citizens, as well as a wide range of environmental impacts. Reducing the paving requirements of a site can lower project costs and reduce stormwater runoff.

Strategies include the following:

- Redevelop sites, such as greyfield or urban infill properties, that are already serviced by roads and other infrastructure.
- Reduce the size of parking spaces, roadways, and driveways.
- Consider shared parking options with neighboring properties.
- Remove paving from the center of turnarounds, cul de sacs, or overly wide roads and replace them with vegetation or other permeable surfaces.
- Design parking so that the front end of automobiles overhang low vegetation or other permeable surfaces.
- Consider two-track driveways and service roads that do not pave the entire area but have a center strip of vegetation or other permeable surface.

■ DESIGN CONSIDERATIONS

Opportunities to reduce paving may be limited by local policies and regulations that dictate the size of roads, parking lots, and other impervious surfaces. In these circumstances, the design team should familiarize themselves with the local requirements and present the environmental and safety benefits to the local authorities in order to obtain a variance.

Permeable Paving

Permeable pavements, also known as pervious or porous pavements, contain voids or pore spaces that allow water to flow through the paving surface to the soil below. In porous asphalt or concrete, pore spaces are created by removing the "fines" (sand-size particles) from the mix and binding angular crushed stone together with asphalt, portland cement, epoxy, or other binders. Other permeable pavements are made from open grid structures or paving units that contain joints filled with either porous aggregate or vegetation (see Figure 5.6).

■ FIGURE 5.6 Permeable pavement.

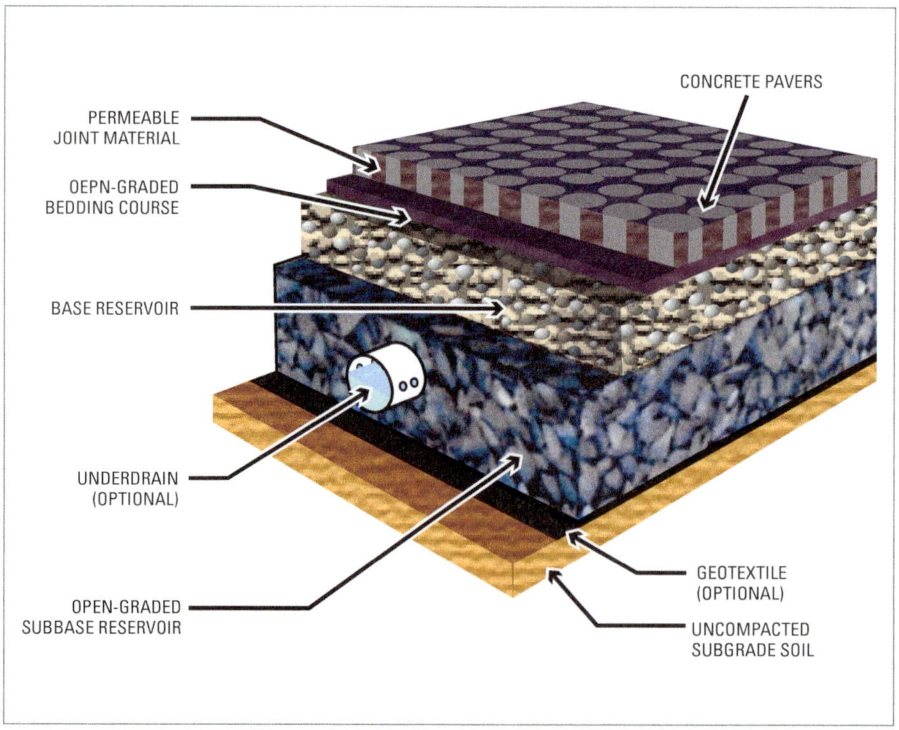

Permeable pavements mimic natural processes within the built environment to reduce runoff and treat stormwater. Actions include:

- **Interception:** Vegetation growing in permeable paving intercepts rainfall.
- **Evapotranspiration:** Water within the pavement can evaporate or be transpired by vegetation.
- **Infiltration:** Permeable paving allows water to come into contact with subsoils.
- **Storage:** Open spaces in the paving material and subbase function as microdetention basins that temporarily store water until it evaporates, percolates into the subsoil, or is transported via a discharge pipe to another location.
- **Bioremediation:** Soil particles, roots, and microorganisms within the paving work to bind and breakdown pollutants.

For sites that require structural paving surfaces, porous pavement can accommodate the required pedestrian and vehicular traffic while also helping to manage storm flows (see Figure 5.7). Infiltration rates vary greatly among the different porous pavement types and often have wide ranges that are dependent upon the construction and maintenance practices employed. Hunt and Bean (2006) tested three materials in similar locations and found the following infiltration rates:

- Concrete grid pavers: 0.99 to 18.8 centimeters per hour
- Permeable interlocking concrete pavers: 100 to 4,000 centimeters per hour
- Pervious concrete: 640 to 6,600 centimeters per hour (Hunt and Bean 2006)

Stormwater that is discharged from pervious paving has gone through an initial filtration and typically has a peak runoff rate that is much lower and later than the peak rainfall. This reduction in runoff volume and delayed release reduces the negative impacts of the runoff.

Porous pavements have been shown to effectively treat pollutants such as oils, nutrients, bacteria, and particulates that are deposited during the course of the pavements' normal use and maintenance (Ferguson 2005). Pollutant removal is accomplished by capturing solid particles and other pollutants and bringing them into contact with vegetation and microorganisms located in the soil and attached to the pavement. Porous paving allows the infiltration and treatment of stormwater to be spread out over the entire paving area, making better use of the land's ability to infiltrate, treat, and store subsurface water.

© CONSERVATION DESIGN FORUM

■ **FIGURE 5.7** Driveway surfaces at Lake Cook Courts are made of porous, interlocking concrete pavers carefully designed to slow, cool, infiltrate, and cleanse rainwater.

■ DESIGN CONSIDERATIONS

In addition to removing pollutants and moderating stormflow, permeable paving can also reduce traffic noise, lower infrastructure requirements, and mitigate the urban heat island. For more on the urban heat island and porous paving, see Chapter 4, "Sustainable Solutions: Air Pollution."

The typical use of permeable pavements is for low-traffic applications such as residential streets, parking lots, emergency vehicle access, driveways, alleys, and trails. Vegetated pavements are often used in areas where seasonal or infrequent use allows plants to regenerate after disturbances and are best suited to climates with regular rainfall.

Permeable pavements can improve driving safety by draining water from the surface of the road and increasing traction. This is particularly valuable in cold climates, where ice accumulation is an issue. Because the hydrologic function of porous pavements is visible, the technology can be an effective educational tool that allows site visitors to see stormwater infiltration and notice the absence of water puddles or runoff.

Special caution should be taken to avoid clogging the pore spaces within the paving material. Sediment, small rocks, and other debris may fill the voids, hindering infiltration and decreasing the utility of the pavement. Landscaping materials such as mulch, topsoil, and sand should not come into contact with porous pavement. For this reason, porous pavements should not receive runoff from unpaved areas. In addition, porous pavement should be avoided in steep subgrade slopes greater than 5 percent due to the additional water movement in the subbase that may cause stabilization issues (Ferguson 2005).

Many permeable pavements require periodic maintenance. Pressure washing or vacuuming the surface may be required if the voids become clogged and the surface no longer infiltrates water as specified. The monitoring and maintenance plan should include strategies for tracking the performance of porous paving and describe the necessary management practices to maintain or improve its function.

■ RESOURCES

Calkins, M. 2009. *Materials for sustainable sites.* Hoboken, NJ: John Wiley & Sons.

Ferguson, B. K. 2005. *Porous pavements.* Boca Raton, FL: Taylor and Francis.

Thompson, J. W., and K. Sorvig. 2000. *Sustainable landscape construction: A guide to green building outdoors.* Washington, DC: Island Press.

Structural Soils

Structural soils are specialized base-course mixtures of open-grade aggregate and soil formulated to support various types of pavement while maintaining favorable growing conditions for vegetation (see Figure 5.8). Void spaces in the mixture are partially filled with soil, with the remaining open space available as a temporary water reservoir. The open-grade aggregate supports traffic load and protects the underlying soil from compaction, which helps maintain healthy root growth (Ferguson 2005).

Structural soils mimic natural processes within the built environment to reduce runoff and treat stormwater. Actions include:

- **Interception:** Vegetation growing in structural soils intercepts rainfall.
- **Evapotranspiration:** Water within the pavement can evaporate or be transpired by vegetation.
- **Infiltration:** Structural soil allows water to come into contact with subsoils.

- **Storage:** The subbase reservoir slows and extends the release of stormwater.
- **Bioremediation:** Soil particles, roots, and microorganisms within the structural soil work to bind and breakdown pollutants.

Stormwater from the immediate vicinity can be directed overland to the structural soil or flow through porous pavement or turf directly into the specialized base course. The amount of water structural soils are able to treat is dependent on the depth of the reservoir. For example, 24 inches (61 cm) of CU-Structural Soil will accommodate 6.25 inches (16 cm) of rain within a twenty-four-hour period, based on a known void space of 26 percent (Haffner and Bassuk 2007). Many different structural soil manufacturers exist. The performance of each system varies and is largely dependent on the materials mix, as well as the installation and maintenance practices. Structural soils are typically designed to drain within forty-eight hours to protect the health of the vegetation and maintain optimal function (Day and Dickinson 2008). Excess water can overflow into a secondary biofiltration treatment feature or storm sewer.

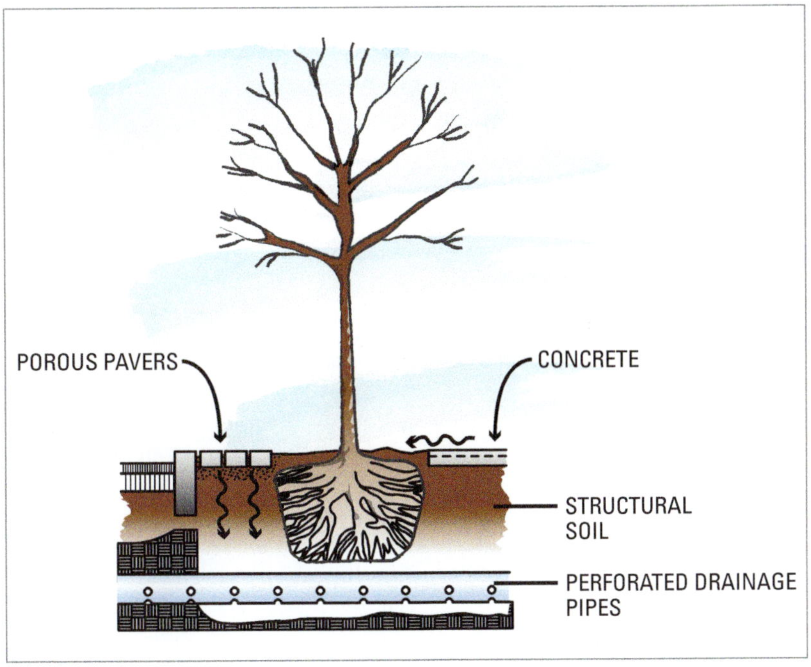

FIGURE 5.8
Structural soils.

DESIGN CONSIDERATIONS

Structural soils allow vegetation—primarily trees—to thrive in areas of the landscape that are typically hostile environments for plants. Increasing vegetative cover provides a variety of benefits for the site and surrounding area, including temperature moderation, improved property values, habitat, and mitigation of the urban heat island.

Structural soils can support both impervious and porous paving in a variety of applications, including sidewalks, driveways, low-use access roads, parking lots, and pedestrian courtyards. The specialized soils encourage deep rooting, which reduces the heaving of sidewalks, driveways, and curbs by tree roots. The depth of the structural soil impacts its ability to support large tree growth, with a 24-inch to 36-inch (61–91 cm) depth being optimum (Haffner and Bassuk 2007).

The monitoring and maintenance plan should include strategies for tracking the performance of the system and describe the necessary management practices. Areas using structural soils will need to be inspected for trash and debris after large storm events. Inlet/outlet pipes should be inspected regularly and cleaned out as needed to prevent clogging.

RESOURCES

Ferguson, B. K. 2005. *Porous pavements.* Boca Raton, FL: Taylor and Francis.

Urban, J. 2008. *Up by roots: Healthy_structural soils and trees in the built environment.* Champaign, IL: International Society of Arboriculture.

Urban Horticulture Institute, Cornell University: CU Structural Soil. http://www.hort.cornell.edu/uhi/

Green Roofs

Green roofs, also known as vegetative, living, or eco-roofs, are specialized roofing systems that support vegetation on both sloped and flat roof surfaces. Vegetative roofs reduce both the volume and rate of stormwater runoff, which can significantly reduce both the spread of pollutants and the demand on stormwater infrastructure (see Figure 5.9).

Green roofs mimic natural systems and reduce stormwater runoff in the following ways:

- **Interception and evapotranspiration:** Vegetation growing in green roofs intercepts and transpires rainfall. Water also evaporates from the soil and growing medium.
- **Storage:** Rainwater is temporarily held by specialized water-holding membrane layers and pore spaces in the soil or growing medium.

FRAUENWOHNEN

■ **FIGURE 5.9** FrauenWohnen extensive green roof vegetated with sedums and native grasses. The green roof is integral to the rainwater management system that results in a zero stormwater runoff site.

The amount of precipitation a green roof can manage is a function of the slope of the roof, the composition and depth of the growing medium, the plant palette, the number and type of roofing layers (e.g., the presence or absence of drainage layers or water-holding membranes), the intensity and frequency of rainfall, and the site's evaporative potential. In general, green roofs are more successful at managing small rain events. Most green roofs can fully absorb rainfall of 0.4 inches (10 mm) or less; however, for 1-inch (28-mm) events, the retention rates can vary from 8 to 43 percent in some systems (Simmons et al. 2008).

■ DESIGN CONSIDERATIONS

In order for a green roof to successfully mitigate urban flooding and water pollution, the system must be specifically designed to address the issue rather than relying on intrinsic benefits believed to be associated with all green roof systems. In addition to stormwater management, green roofs can also provide habitat, increase aesthetic value, and help mitigate the urban heat island effect. A more detailed discussion of green roofs and these benefits can be found in Chapter 4, "Sustainable Solutions: Air Pollution."

Certain green roof growing media and maintenance practices, such as fertilizer and pesticide application, can pollute stormwater runoff. To avoid contamination, give special attention to the characteristics and components of the growing medium and its potential to bind and retain or leach pollutants (Köhler and Schmidt 2003). In addition, the site maintenance plan should outline the maintenance practices that need to be implemented or avoided in order to protect water quality. Retaining and treating pollutants on-site can also be accomplished by making the green roof part of a stormwater treatment train, where runoff from the roof is conveyed to a second biofiltration area such as a bioswale or rain garden.

■ CASE STUDY

LAKE COOK COURTS

PROJECT TYPE: Multifamily residential

LOCATION: Highland Park, Illinois

SIZE: 4.4 acres (1.8 hectares)

COMPLETION DATE: 2008

HIGHLIGHTED SUSTAINABLE PRACTICES:

- Redevelopment of an urban infill site
- Removal of invasive species
- Restored habitat
- Minimized impervious surfaces
- Green stormwater infrastructure
- All residents have visual and physical access to nature
- Post-development maintenance and stewardship plan

■ **FIGURE 5.10**
Lake Cook Courts schematic plan.

© CONSERVATION DESIGN FORUM

THE SITE: The urban infill site was the location of several single-family homes, one of which included the original 125-year-old farmhouse for the area. The property is located on a regional arterial roadway and is surrounded by single-family homes and a regional mall. The landscape was primarily turfgrass and invasive trees, shrubs, and herbaceous species.

Design Overview

The plan for Lake Cook Courts clusters seventeen homes along a narrow brick lane and parking court (see Figure 5.10). The layout provides a range of home sizes and includes the preservation and restoration of an 1885 farmhouse as well as two affordable homes. Although clustered very close together, the site plan maximizes privacy and long views to the community landscape through carefully orchestrated window placement and staggered setbacks. Each home has a private outdoor patio space designed to complement the community landscape, and terraces have been carefully situated to provide personal space for relaxing and entertaining. Views from every window look out on lush native prairie plantings, while rooftop terraces provide another indoor/outdoor experience (see Figure 5.11).

continues

LAKE COOK COURTS *(CONTINUED)*

■ **FIGURE 5.11**
Homes are clustered together while maximizing privacy and views to community green space.

The landscape around the homes combines low-input ornamental perennials and naturalized plantings and includes a series of rain gardens that are part of the integrated rainwater management system. Residents share a small walking path that meanders through the common space. Restored habitat attracts birds and butterflies, and provides an interesting community setting that changes with the seasons (see Figure 5.12). Large oaks and other native trees were carefully preserved, while invasive understory trees and shrubs were removed. Invasive plants were managed during establishment of the newly planted landscape as identified in the maintenance and stewardship plan developed for the project.

Stormwater is considered a resource and is managed throughout the site with a series of integrated features. Lanes and driveways are made of porous, interlocking concrete pavers engineered to slow, cool, infiltrate, and cleanse rainwater. Bioswales further treat and utilize rainwater, avoiding downstream discharge. Rainwater is visibly directed through scuppers and rain chains, and then to a small gravel swale between each of the homes. It then flows to rain gardens incorporated into the ornamental landscape at the rear of the homes. Any surplus water that isn't absorbed can then overflow through a shallow swale to a large depressional area in the common space that has been established with native prairie. The entire system will retain up to 2 inches (5.1 cm) of rainfall on-site with no surface water discharge; only during a very extreme rain event does any rainwater leave the site, and then only after it has been cooled and filtered by the landscape.

LAKE COOK COURTS *(CONTINUED)*

PROJECT TEAM

- **LANDSCAPE ARCHITECTURE AND WATER RESOURCE ENGINEERING:**
  Conservation Design Forum
  David Yocca, FASLA, LEED AP
  Tom Price, PE
  http://www.cdfinc.com/

- **ARCHITECT:**
  Yas Architecture
  Stephen Yas, AIA
  http://www.cdfinc.com/

© CONSERVATION DESIGN FORUM

■ **FIGURE 5.12**
Purple coneflower (*Echinacea purpurea*), black-eyed Susan (*Rudbeckia hirta*) and white coneflower (*Echinacea purpurea* 'Alba') in the community green space.

Downspout Disconnection

In many urban areas, downspouts direct water from rooftops to sewer systems, bypassing the surrounding landscape. This practice robs the site of a valuable water resource and increases stormwater volume, downstream flooding, and the likelihood of combined sewer overflows. In most circumstances, downspouts can be disconnected from existing standpipes and redirected into landscaped areas, lawns, or other stormwater management features (see Figure 5.13). Downspouts can also flow into cisterns or rain barrels, making the water available for reuse at a later date.

Many communities provide incentives for disconnecting downspouts. One example is the city of Portland, Oregon, which has worked with households and small commercial buildings to redirect more than 1.2 billion gallons (4.5 million kL) of stormwater to rain gardens or cisterns, resulting in a significant reduction of water pollution and combined sewer overflows (City of Portland, Oregon 2010).

FRAUENWOHNEN

■ **FIGURE 5.13**
Drainage pipes at FrauenWohnen release rainwater into gravel-filled channels that slowly transport water to a vegetated swale.

■ DESIGN CONSIDERATIONS

When considering downspout disconnection, the first step is to determine whether the water will go to a cistern or to the landscape. If the landscape is chosen, designers must then determine where the water will be redirected. As a general rule of thumb, the receiving area should be at least 10 percent of the roof area that drains to the downspout. Sites with limited space can direct runoff to rainwater harvesting cisterns. Septic fields and other areas not suitable for stormwater infiltration should be avoided. Other specifications include:*

- Six feet (1.8 m) from basement walls
- Two feet (0.6 m) from crawl spaces and concrete slabs
- Five feet (1.5 m) from the property line
- Three feet (0.9 m) from any sidewalks
- Ten feet (3 m) from a retaining wall

* City of Portland 2009

■ RESOURCES

City of Portland, Oregon. Environmental Services. 2009. "How to manage stormwater: downspout disconnection."

Water Environment Research Foundation. "Downspout disconnection.": http://www.werf.org

Protect and Restore Soil Health

Healthy soils are key to a sustainable site, but their importance is often underestimated. Properly functioning soils provide a variety of environmental and economic benefits, including:

- Supporting the growth of vegetation
- Infiltrating precipitation, storing water, and reducing flooding
- Treating and filtering water pollutants
- Sequestering atmospheric carbon
- Providing habitat for a variety of plants and animals

Compaction and soil organic matter are two major factors influencing soil health and its ability to absorb, retain, and cleanse water. Both are discussed below in greater detail.

Soil Compaction

Weight from a single intense force or small repeated forces pushes soil particles together, causing them to compact. Compacted soils have reduced macro and micro pore space, which results in restricted root growth, reduced infiltration rates, and decreased biological activity. Overcompacted soils greatly limit plant growth and the ability of a site to absorb and cleanse stormwater (see Figure 5.14).

HEATHER VENHAUS

■ FIGURE 5.14
Severe erosion and tree mortality caused by soil compaction from pedestrian traffic.

Common causes of soil compaction include:

- Construction and maintenance equipment
- Parking or driving on portions of the site not designated for vehicular traffic
- Repeated pedestrian and animal traffic
- Compressing or working soils while wet
- Rainfall on bare soils
- Low levels of organic matter

Soil compaction is defined by an increase in bulk density, which is the dry weight of soil divided by its volume, and is commonly expressed in grams per cubic centimeter (g/cm^3) or megagrams per cubic meter (Mg/m^3). Soils have varying densities at which they will no longer remain stable or support root growth, commonly known as the "growth limiting" or "maximum allowable" bulk density. A range of acceptable bulk densities, expressing the lower and upper limits (based on the soil texture), should be determined by the project team (see Figure 5.15). Knowing the bulk density range is important for identifying the vegetated areas of the site that will need to be restored in order to improve growing conditions.

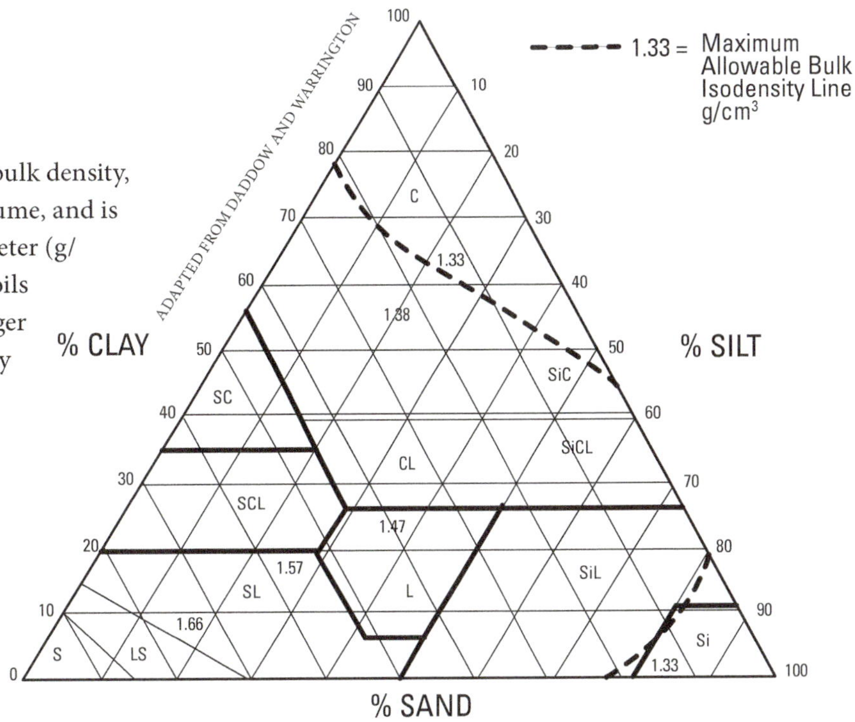

■ FIGURE 5.15
Maximum bulk densities recommended by SITES.

Potential signs of soil compaction include:

- Water ponding
- Surface-water runoff
- Soils that are bare and not supporting vegetation
- Shallow tree-rooting
- Stunted vegetation

Many of these features may also be indicators of other site conditions and should not be used in isolation to determine whether soils are overly compacted. To confirm compaction levels, design teams can use tools such as cone penetrometers (ASTM D3441) or bulk density tests (ASTM D4564).

Strategies for Preventing or Minimizing Soil Compaction

Protecting soils from compaction can save significant time and money over the life of a project. Unnecessary plant replacement costs, drainage issues, and erosion-control measures can all be avoided. Opportunities to be good stewards of soil health exist at each phase of the project. Strategies include the following:

■ DESIGN

- Map areas of degraded soil conditions, such as overcompaction, erosion, or contamination.
- To the greatest extent practicable, avoid grading, vegetation removal, or other disturbances of healthy soils. Locate site features that require soil disturbance on areas of existing degraded soils.
- Develop a soil preservation plan outlining areas not to be disturbed. Fence these areas and clearly communicate the plan and its importance to construction personnel.

■ CONSTRUCTION AND MAINTENANCE

- Enforce tight limits of disturbance during the construction process. The Sustainable Sites Initiative (2009) recommends that disturbance be limited to 40 feet (12 m) beyond the building perimeter; 10 feet (3 m) beyond surface walkways, patios, parking, and utilities that are less than 12 inches (30 cm) in diameter; 15 feet (6 m) beyond primary roadways, curbs, and main utility branch trenches; and 25 feet (8 m) beyond constructed areas with permeable surfaces, such as stormwater detention facilities and recreation fields.
- Designate areas for on-site parking, equipment, and material storage. Prioritize the use of areas that are already degraded or will be compacted in order to support a future use, such as patio, driveway, or building site. Explore options for using existing roads or parking areas adjacent to the site for access and storage.
- In areas where compaction cannot be avoided, carefully harvest and store the topsoil for reuse.
- During construction, spread thick layers of mulch over soil that may receive occasional traffic. Sheets of plywood can be added on top of the mulch to help spread the weight in areas of repeated traffic. Geogrid or other geotextiles can be placed under the mulch to provide an additional weight dispersal mechanism.
- Avoid working the soil when wet. The soil should be considered too wet when it is moist enough to stick to your hand and make impressions of your fingers when squeezed (Urban 2008).

- Avoid bare soils. Cover soils with mulch or vegetation.
- Use the lightest equipment possible and consider the surface area over which the weight will be distributed. Tracked vehicles and equipment with low-inflation rubber tires spread their weight, resulting in a much lower compacting force (Thompson and Sorvig 2000).
- Use caution when turning equipment in order to avoid site damage. Select equipment that can easily navigate the landscape and does the least damage to the soil and vegetation. Tracked vehicles can disturb large areas when turning, due to the entire track skidding over the soil surface.

Strategies for Restoring Overly Compacted Soils

Soils with limited infiltration capacity and bulk densities that restrict root growth should be confined to areas of the site that require compacted soils for structural support, such as the subbase for buildings, roads, and sidewalks. All other soils that are intended to be revegetated should be restored as necessary to sustainably support the selected vegetation. Efforts to rehabilitate compacted soils typically involve three steps:

1. Break apart compacted soils through practices such as tilling or subsoiling. Work cautiously under and around existing vegetation to avoid damaging root systems. An arborist can provide guidance on the most appropriate methods, which may include air-excavating tools, vertical mulching, or radial trenching.
2. Incorporate compost or mineral amendments into the soil. Compost is preferable in most applications because of the many soil-health benefits it provides.
3. Protect soils from recompaction. Revegetate and limit pedestrian and vehicular traffic.

Because soil conditions vary between and within sites, restoration strategies must be tailored to the site's specific characteristics and requirements. Consult soil and vegetation experts to ensure that soil modifications address site conditions and create long-lasting and sustainable outcomes.

Organic Matter

Organic matter originates from living organisms such as leaves, roots, worms, and insects and can be made up of both living and dead materials. Soil microorganisms, which rely on organic matter as a food source, break down the materials to make nutrients available to vegetation, improve soil aggregation, and create humus, which can absorb and hold large quantities of water. Organic matter improves the drainage rates of clay and silt soils and enhances the water-holding capacity of sandy soils.

Organic matter, in terms of volume, is a relatively minor component of the soil; however, its influence on soil function is quite large. Soil organic matter has the following benefits:

- Promotes good soil structure
- Reduces soil compaction
- Provides a food source for soil microorganisms
- Improves infiltration and air movement through the soil
- Increases water storage
- Provides nitrogen and other nutrients needed by vegetation
- Removes or binds pollutants

Healthy topsoil in most temperate regions contains between 3 and 5 percent organic matter; however, levels can vary between ecosystems. For example, wetlands typically have very high organic matter content, while native desert soils often contain less than 1 percent. Urban soils commonly suffer from lower percentages than the surrounding native ecosystems due to land construction and maintenance practices that remove vegetation from the site and compact soils (Urban 2008).

When making soil management decisions, identify a reference soil to assist in determining the appropriate organic matter content and other soil conditions. Reference soils should support vegetation and ecological functions similar to those intended for the site. Design teams can use existing portions of the project area as a reference or look to healthy and functioning ecosystems in surrounding areas for guidance.

Strategies for Maintaining Appropriate Levels of Soil Organic Matter

Soil organisms break down organic matter into nutrients and other substances beneficial to vegetation. Because organic matter continually decomposes, it must be regularly replenished. Conventional maintenance practices, such as raking and bagging lawn clippings and leaves, rob the soil of its natural source of organic matter. Sustainable sites mimic natural ecosystems by regularly replenishing the organic matter content of the soil. Strategies include:

- **Avoiding soil disturbance:** Tillage and other forms of soil disturbance promote the loss of organic matter by speeding decomposition.
- **Leaving discarded and decaying plant materials on-site:** Allow leaves, stems, and other materials to decay on the landscape and become naturally incorporated into the soil.
- **Providing regular inputs of organic materials:** Top-dress planting beds with organic materials such as shredded leaves, straw, or compost.
- **Maintaining or increasing vegetative cover:** Plants deposit organic matter onto the soil surface and within the soil profile.

As a reference, the Sustainable Sites Initiative recommends that a minimum of the top 12 inches (30.5 cm) of soil contain at least 3 percent organic matter or organic matter levels comparable to similar surrounding native landscapes serving as a reference site.

■ DESIGN CONSIDERATIONS

Managing organic materials on-site not only reduces the amount of waste leaving the landscape but also reduces transportation and disposal costs, saves natural resources, and prevents pollution. Options for reusing organic materials from both the landscape and buildings should be explored early in the design process to allow adequate space to create and store mulch and compost.

Increasing the organic matter content of a soil can require large amounts of compost or other organic amendments and is typically accomplished slowly over time. The beneficial effects of increasing organic matter can begin long before soil levels rise; however, the gains can easily be reversed by returning to conventional construction and management practices that disturb the soil and remove vegetation from the site.

One type of organic soil amendment that should be avoided is sphagnum peat. Peat is a nonrenewable resource that sequesters significant amounts of carbon and is often transported great distances. Because the embodied carbon of peat is so great, it is difficult for sites to offset the initial environmental damage caused by the extraction of the material.

As with all soil amendments, the application of compost should be carefully managed. Though it is a natural product, compost can leach nutrients, causing pollution of surface water and groundwater. When applying compost in large amounts or near sensitive environmental areas, take steps to prevent runoff and avoid contamination to receiving water bodies. The potential for water pollution is particularly great in landscape features that are intended to manage stormwater, such as green roofs or biofiltration areas. Therefore, the use of compost in these systems is often debated, and special consideration should be given to the organic matter type and its potential for water pollution.

■ RESOURCES

Calkins, M., ed. 2011. *Sustainable sites handbook.* Hoboken, NJ: John Wiley & Sons.

Craul, T., and P. Craul. 2006. *Soil design protocols for landscape architects and contractors.* Hoboken, NJ: John Wiley & Sons.

Urban, J. 2008. *Up by roots: Healthy structural soils and trees in the built environment.* Champaign, IL: International Society of Arboriculture.

Compost is a readily available source of organic matter and available nutrients. It is created by the controlled biological decomposition of organic materials such as leaves, branches, and food scraps diverted from the waste stream (U.S. Composting Council 2008). Compost provides the following benefits:**

- Improves soil structure
- Supports soil microorganisms
- Decreases soil compaction
- Improves the water-holding capacity of soil and reduces irrigation demands
- Provides nutrients and improves soil fertility
- Binds heavy metals and degrades, or in some cases completely eliminates, wood preservatives, petroleum products, pesticides, and both chlorinated and nonchlorinated hydrocarbons in contaminated soils
- Diverts organic materials from the waste stream, extending the life of municipal landfills
- Improves the health of vegetation, increasing the aesthetic quality of the site

Compost works best when it is tailor-made or specially designed for the specific use and soil type. Technical parameters such as maturity, stability, pH level, density, particle size, moisture, salinity, and organic content can all be adjusted to meet the needs of the site (U.S. EPA 1997). The U.S. Composting Council has developed standards for compost that can be referenced when developing compost specifications.

** U.S. EPA 1997

Strategies for Restoring Soil Organic Matter

Restoring appropriate organic matter content is a relatively simple and cost-effective practice (Urban 2008). An option for soils that need only minor improvements is to continually top-dress planting beds with several inches of compost or mulch. Earthworms and other soil fauna will gradually work the vegetative materials into the soil, building organic matter content over time. A second and more direct approach is to till a generous amount of mature and stable compost into the top 6 to 18 inches (15 to 46 cm) of the soil. The specific quantity of compost required depends on the existing soil type and conditions as well as the organic matter content of the compost. Compost manufacturers should be able to provide independent third-party test results indicating the organic matter content, pH, nutrient availability, soluble salts, and other important factors.

It is possible to add too much organic matter to soils; this can cause drainage problems, nutrient loading, settling, and other issues. In addition, dramatically increasing the organic matter content of soils that are naturally low in organic matter can essentially confine plant roots to the amended area. To avoid this problematic scenario, the organic matter content of topsoil and subsoil layers should match the reference soil conditions.

Soil Microorganisms

One teaspoon of healthy soil contains millions of such beneficial microorganisms as bacteria, fungi, and earthworms (Ingham 2000). Air, water, and organic matter within the soil support the microorganisms, which in turn support healthy plant growth, nutrient cycling, pollutant removal, and the enhancement of soil structure (see Figure 5.16). The diversity and abundance of microorganisms is directly related to the organic matter content of the soil.

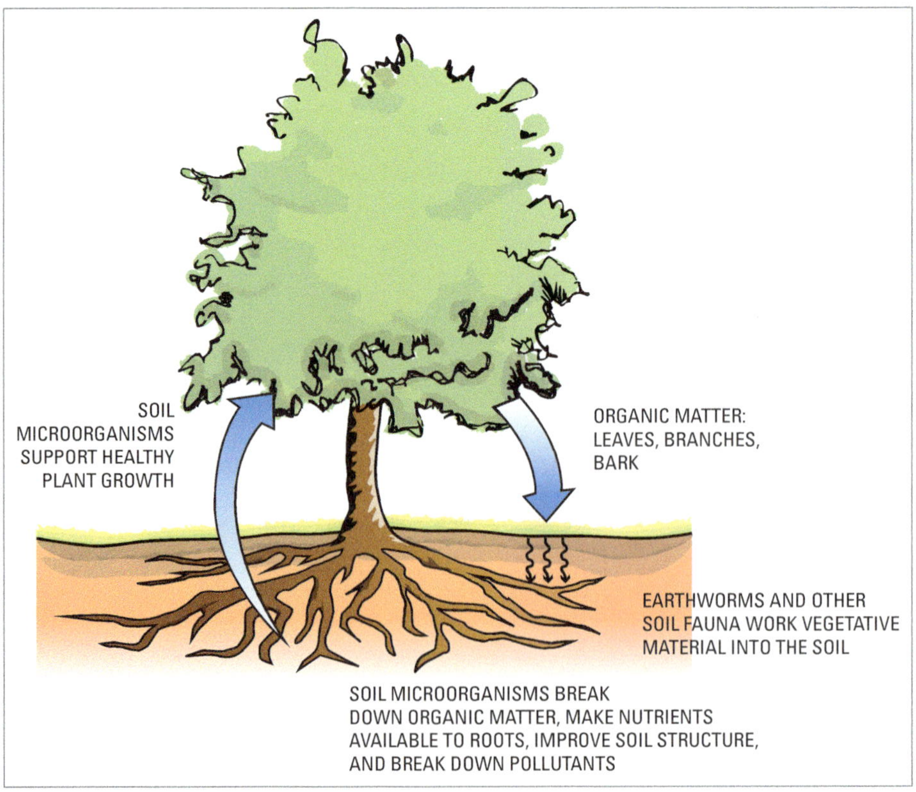

HEATHER VENHAUS

■ FIGURE 5.16 Symbiotic relationship between vegetation and soil microorganisms.

Soil organisms require air and regular inputs of organic matter. Strategies for protecting and encouraging soil organisms include:

- Limiting soil disturbance and tillage
- Reducing compaction to allow air and water movement through the soil
- Maintaining appropriate soil organic matter levels
- Avoiding pesticide use that may harm soil biota
- Providing a variety of food sources by maintaining a diverse plant palette
- Avoiding bare soil by maintaining vegetative cover, mulch, or plant materials such as leaves

Bioremediation

Naturally occurring soil microorganisms and their enzymes can break down or immobilize a variety of contaminants, such as wood preservatives, pesticides, and petroleum products. The use of microbes to clean contaminated soil or water is known as bioremediation. Stormwater management practices that use soil and vegetation to treat runoff, such as rain gardens or bioswales, capitalize on the inherent water-cleansing benefits provided by soil microbes and are an example of on-site bioremediation (see Figure 5.17).

When the necessary microorganisms are not present, specialized microbes can be introduced to degrade contaminants. Bioremediation is commonly employed in natural resource extraction industries such as petroleum and mining. Because bioremediation uses resources available on-site to clean up contamination, it is typically more cost-effective than chemical treatment processes, and less

HEATHER VENHAUS

■ **FIGURE 5.17**
Stormwater from the staff parking lot at the Lower Colorado River Authority Redbud Cener flows overland into vegetated terraces that infiltrate and cleanse the runoff. Design strategies that use soil and vegetation to treat runoff capitalize on the inherent water-cleansing benefits provided by soil microbes and are an example of on-site bioremediation.

damaging to the environment (U.S. EPA 2001a). Bioremediation efforts can effectively clean up many types of pollutants but is largely unfeasible for projects with high concentrations of substances such as lead, salts, or cadmium, which are toxic to most microorganisms.

■ RESOURCES

Calkins, M., ed. 2011. *Sustainable sites handbook.* Hoboken, NJ: John Wiley & Sons.

Craul, T., and P. Craul. 2006. *Soil design protocols for landscape architects and contractors.* Hoboken, NJ: John Wiley & Sons.

Ingham, E., A. Moldenke, and C. Edwards. 2000. *Soil biology primer.* Ankeny, IA: Soil and Water Conservation.

Urban, J. 2008. *Up by roots: Healthy structural soils and trees in the built environment.* Champaign, IL: International Society of Arboriculture.

Increase Vegetative Cover

Vegetation controls stormwater at the source by intercepting rainfall and encouraging infiltration. A portion of each rain event is intercepted by vegetation and temporarily held on the leaves, stems, and branches until the water evaporates back into the atmosphere or is gradually released to the surface below.

The percentage of annual rainfall intercepted by vegetation is strongly influenced by rainfall patterns and the plant type, size, and foliation period (Xiao et al. 2000). Trees intercept significant quantities of precipitation because of their large leaf area. On average, broadleaf evergreen trees intercept the most rainfall, followed by conifer and broadleaf deciduous (Xiao and McPherson 2002). Planting street trees has become a common stormwater management practice in many urban areas. In New York City, it is estimated that street trees intercept 890.6 million gallons (3.4 million kl) of stormwater each year, resulting in reduced infrastructure and water pollution costs and ultimately providing an annual benefit to the city of $35.6 million (Peper et al. 2007).

■ FIGURE 5.18 The role of vegetation in stormwater management.

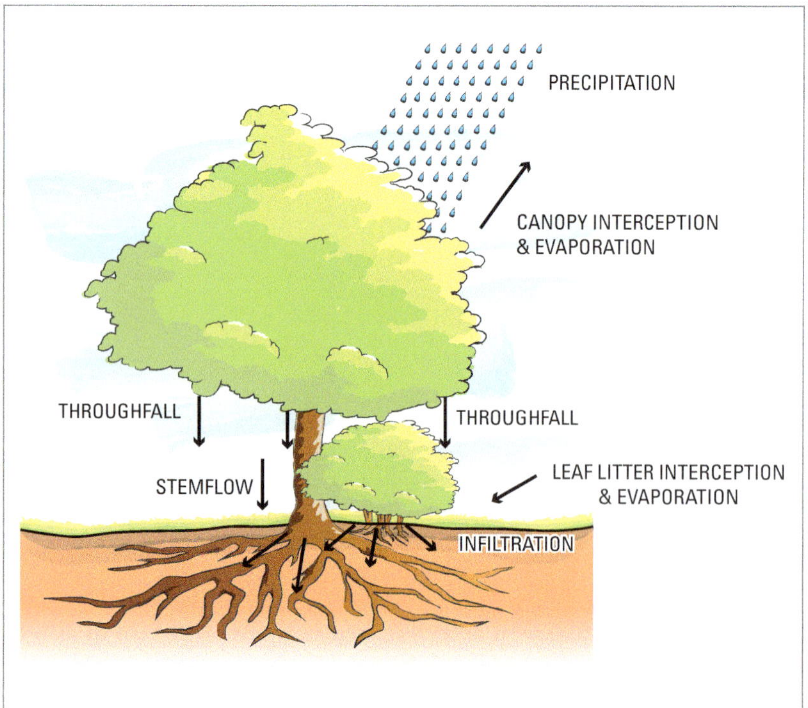

In addition to interception, vegetation also reduces runoff and water pollution by improving infiltration. Plants increase the infiltration rate and water-holding capacity of soil by reducing compaction, supporting healthy soil structure, and increasing soil porosity. Vegetation also draws down soil water content between rain events, which increases the amount of water the soil can hold during the next event (see Figure 5.18).

Strategies for Increasing Vegetative Cover

Planting schemes that provide multiple layers of vegetation—groundcover, shrub, midstory, and overstory—can greatly increase vegetative biomass on a site, especially when compared to conventional landscapes that primarily have two layers: groundcover and upper story. Design solutions that allow vegetation to be grown in or over conventional impervious surfaces, such as green roofs, green walls, structural soil, and living walls, also increase the vegetated area of a site and in some circumstances can completely cover all impervious surfaces.

■ DESIGN CONSIDERATIONS

There are limits to which vegetation can be sustainably increased on a site. The vegetated area should be determined by the availability of precipitation, nonpotable water, nutrients, and other resources that can be sustainably provided by a site. Through on-site composting and the creative reuse of alternative nonpotable water resources such as stormwater, air conditioner condensate, and greywater, urban landscapes can support an abundance of vegetation. Reusing these resources on-site has the added benefit of avoiding unnecessary economic and environmental costs related to the transportation, treatment, and disposal of the materials.

Vegetation is a key component of a sustainable site. It plays an integral role in the earth's major biogeochemical cycles, such as the hydrologic, nitrogen, and carbon cycles, and provides a variety of ecosystem services, including:

- Regulates and moderates local and global climate. Vegetation helps maintain a balance of atmospheric gases, provides oxygen, and sequesters greenhouse gases. Local temperatures are regulated through evapotranspiration, shading, and windbreaks.
- Provides food and renewable nonfood products such as wood, fiber, oils, and organic matter.
- Cleans air and water. Vegetation absorbs, sequesters, and breaks down pollutants in air and water.
- Provides erosion control by intercepting rainfall, increasing infiltration, and helping hold soil together.
- Provides habitat. Vegetation provides refuge, breeding and nursery habitat for wildlife.
- Provides medicinal resources. Plants contribute to many chemical compounds used directly or modeled to create pharmaceuticals.

Vegetation and Pollutant Removal

Many pollutants attach to sediment or other suspended particles and are transported by stormwater runoff. As the runoff moves across the landscape, vegetation reduces the velocity of the water and captures pollutants by allowing sediment to settle out of suspension. Plants absorb, accumulate, and in some cases break down pollutants such as heavy metals, explosives, pesticides, and other toxic materials in soil and groundwater. Plant roots uptake these harmful substances along with water and nutrients. Soil microorganisms associated with plant roots or other organic matter can also break down and immobilize pollutants. In addition, vegetation can prevent wind, rain, and groundwater from transporting the pollution to other sites.

The intentional use of plants to remediate soil pollution is called phytoremediation, a strategy most commonly used where soil or water are known to be polluted and specific contaminates are being targeted (see Figure 5.19).

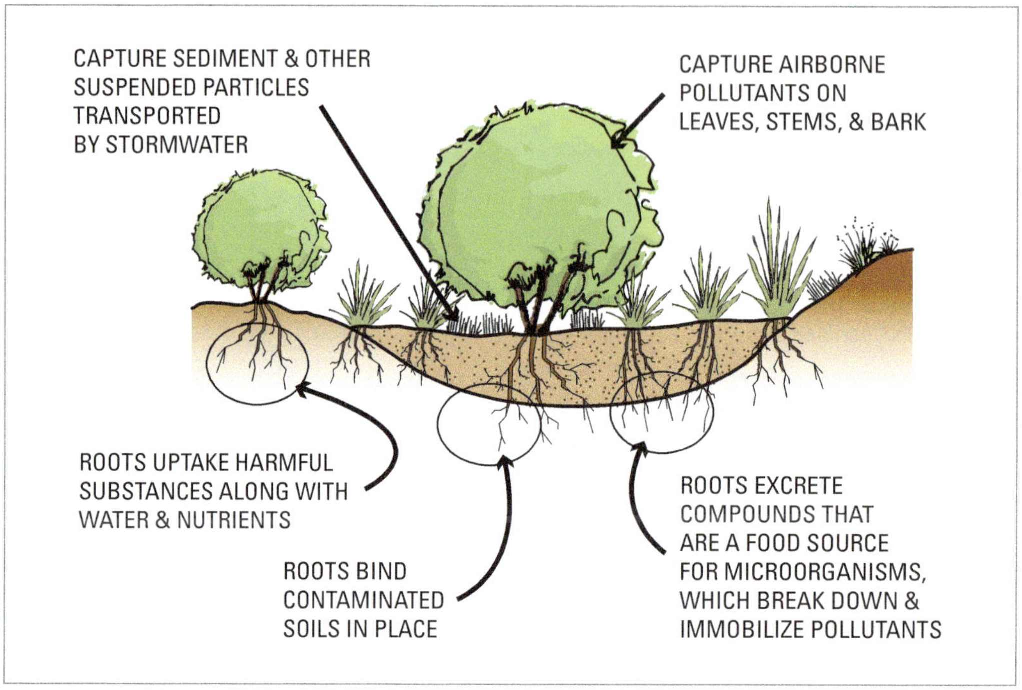

■ **FIGURE 5.19** Phytoremediation.

Phytoremediation is best used on sites with low to medium concentrations of pollutants located in the upper soil layers (U.S. EPA 2001b). Phytoremediation is a new and innovative field in which there is still much work to be done. Projects teams interested in phytoremediation should consult a specialist to help evaluate the utility of this technique for their particular circumstance.

■ RESOURCES

Dunnett, N., and N. Kingsbury. 2004. *Planting green roofs and living walls.* Portland, OR: Timber Press.

Green Values Stormwater Toolbox. National Stormwater Management Calculator.

http://greenvalues.cnt.org/national/calculator.php

U.S. EPA. Office of Soils Waste and Emergency Response. 1999. Phytoremedation Resource Guide. Washington, DC: US EPA.

CASE STUDY

ARKADIEN ASPERG

PROJECT TYPE: Multifamily residential

LOCATION: Asperg, Germany

SIZE: 3.7 acres (1.5 hectares)

COMPLETION DATE: 2002

CLIENT: Strenger Bauen & Wohnen GmbH

HIGHLIGHTED SUSTAINABLE PRACTICES:

- Rainwater harvesting and reuse
- Encourages social interaction and physical play
- All residents have views and access to green space
- Stormwater management features beautifully integrated into the site in a way that encourages interaction with the nature
- Manages the two-year storm event on-site

THE SITE: A newly developed site in Stuttgart, Germany

© ATELIER DREISEITL

FIGURE 5.20
Arkadien Asperg site plan.

Design Overview

Arkadien Asperg is an urban village with a verdant, garden-city ambience nestled within the congested conurbation of Stuttgart. The Arcadian concept includes a distinct design quality for open space, with small neighborhood squares and a central plaza, inviting people of all ages to enjoy the social community. The development contains sixty dwellings per hectare (2.5 acres) and includes low-income housing. A mix of semipublic and private spaces harmonizes a wide variety of passive solar housing types. All dwellings include either gardens or generous balconies that serve as seductive sun-traps (see Figure 5.20).

DIETER GRAU / © ATELIER DREISEITL

FIGURE 5.21
A rainwater-fed stream with natural banks flows through the community and into public gathering spaces where site users can freely interact with the water.

continues

ARKADIEN ASPERG *(CONTINUED)*

The stream is an enjoyable highlight of the extensive stormwater management system, which includes green roofs, various permeable surfaces, and fourteen decentralized cisterns (see Figure 5.22). The main cistern holds 15,850 gallons (60m^3) and is used to supply the stream, while a network of smaller cisterns are distributed among individual houses for toilet flushing, irrigation, and household laundry. Vegetated stonewalls, natural stone, wooden structures, and generous informal planting accompany the stormwater features, lending garden-city flair to the housing estate.

PROJECT TEAM

- **LANDSCAPE ARCHITECTS**
  Atelier Dreiseitl
  www.dreiseitl.com
- **ARCHITECT**
  Joachim Eble Architekture
  www.eble-architektur.de
- **CONSTRUCTION SUPERVISION**
  Ikarus Architekten
- **ENERGY CONCEPT**
  Steinbeiss Gruppe
  http://www.stz-egs.de
- **COLOR CONCEPT**
  Lasuveda
  http://www.lasuveda.de
- **DEVELOPER**
  Strenger
  www.strenger.de

DIETER GRAU / © ATELIER DREISEITL

FIGURE 5.22
The rainwater-fed stream is a highlight of the extensive stormwater management system, which includes green roofs, various permeable surfaces, and fourteen decentralized cisterns.

Slow Stormwater Runoff and Improve Infiltration

Stormwater is a valuable resource that is wasted when it is quickly transported away and not used to support the water needs of the project. Throughout the design process, the integrated design team should seek opportunities to cleanse and reuse stormwater on-site. When soil and vegetation capture and cleanse stormwater, numerous environmental and economic benefits can be realized, including improved aesthetic quality, habitat, groundwater recharge, and air- and water-pollutant removal. This contrasts with conventional stormwater management strategies that typically have one purpose—to move water—and limited additional benefits. Strategies that accomplish multiple objectives make better use of resources and are a wiser, more sustainable choice.

Site strategies for slowing runoff and encouraging infiltration include:

- Increasing vegetative cover
- Restoring degraded and compacted soils
- Incorporating biofiltration features into the site
- Harvesting rainwater

Stormwater Treatment Train

In many cases, a single stormwater management strategy may not be sufficient to manage and treat all runoff. In these situations, a series of strategies can be linked together to form a stormwater treatment train. In the train, overflow from one strategy moves into the next, creating an additional opportunity for the water to be fully managed and treated on-site (see Figure 5.23).

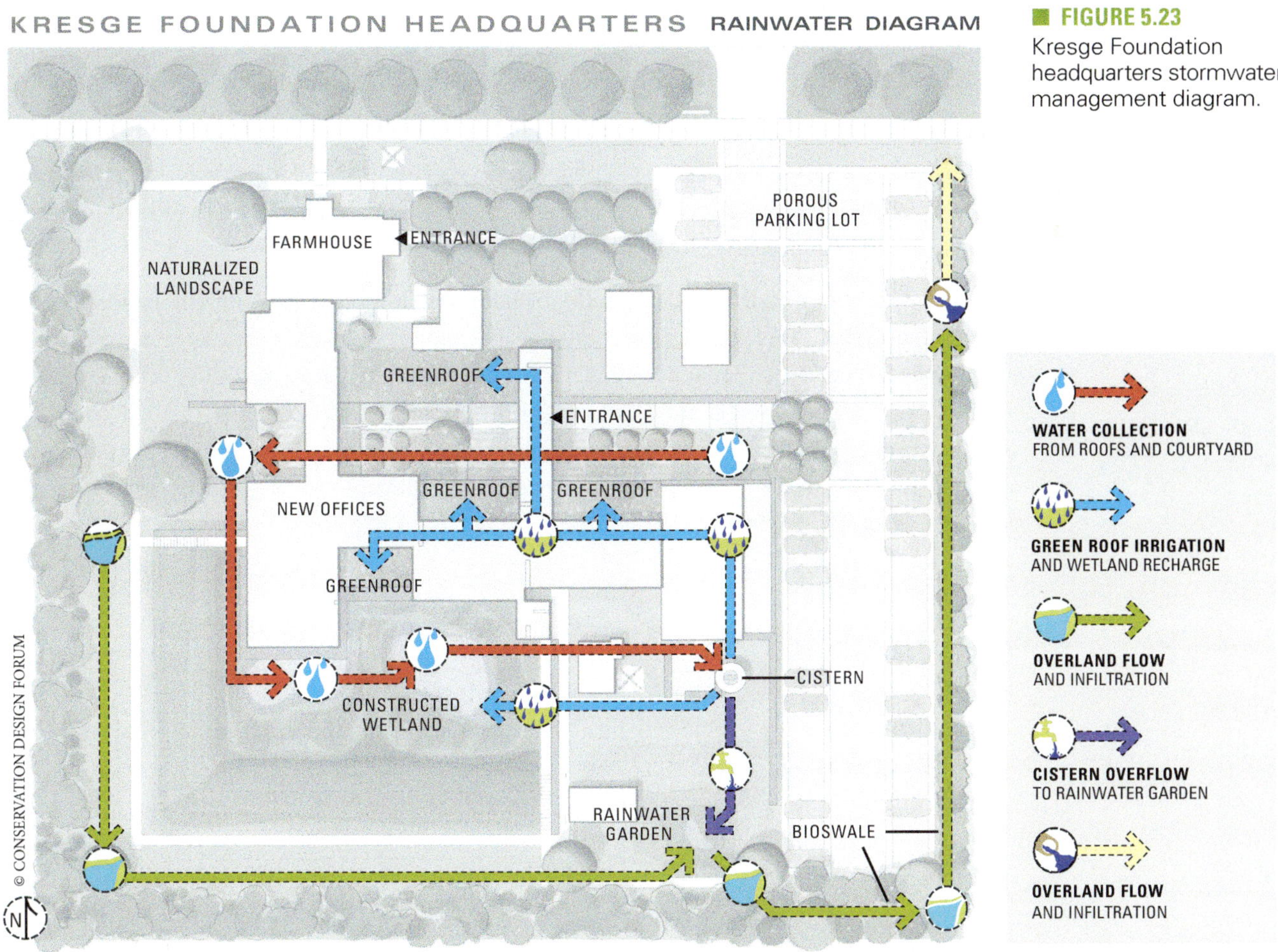

FIGURE 5.23
Kresge Foundation headquarters stormwater management diagram.

For example, runoff from a green roof may flow into a stormwater planter, then overflow into a bioswale that releases into a rain garden (Dunnett and Kingsbury 2004). Creatively managing stormwater in this fashion not only adds interest to the landscape but also provides educational opportunities by revealing ecological processes and reducing the environmental and economic damage associated with urban runoff.

When to Avoid Stormwater Infiltration

Low-impact development or green infrastructure practices that manage stormwater on-site are possible on almost any landscape; however, direct infiltration of stormwater is not always advisable. Under the wrong circumstances, stormwater can pollute groundwater and compromise the structural stability of buildings and other site features (Ferguson 2005). The following site characteristics should be carefully considered prior to using a stormwater management practice that encourages significant infiltration:

- High water tables that may come into contact with stormwater that has not been adequately treated (Prince George's County 2007)
- Shallow confining layers such as bedrock or a clay pan that could prevent deep groundwater recharge
- Areas within 10 feet (3 m) of buildings that may cause potential structural damage (Ferguson 2005)
- Low-permeability soils as indicated by puddling or ponding (Ferguson 2005)
- Septic tanks or leach fields
- Soils with high permeability, such as those over shallow or exposed karst geology that may not adequately treat stormwater and cannot be amended to provide the necessary treatment
- Polluted soils that may leach contaminants (Ferguson 2005)
- Areas where compacted soils are required to support hardscape, structures, and other site features
- Stormwater hot spots where land use may generate runoff that is especially contaminated. Such areas include fueling stations, commercial nurseries, vehicle service and maintenance areas, and auto recycling facilities (Prince George's County 2007).

In circumstances where direct infiltration is inadvisable, stormwater can still be managed on-site using strategies that capture and divert water away from problematic areas. For example, rainwater can be harvested in cisterns and made available for reuse within buildings or in the other portions of the landscape where infiltration is not a concern. Similarly, blue roofs can temporarily hold the rainwater until it evaporates or can use it to irrigate the site. Other options include underdrain systems that allow some infiltration to occur but detour water at certain a soil depth before it interferes with groundwater or reaches contaminated soil layers.

Biofiltration Areas

Biofiltration areas are site features that use plants, soils, mulch, and microbes to slow and treat stormwater runoff. This decentralized stormwater management practice is modeled after natural ecosystems and has been shown to effectively reduce heavy metals, nutrients, harmful bacteria, water temperatures, and other pollutants (Prince George's County 2007). Biofiltration areas can be used in both urban and rural sites and designed to blend with the surroundings and function as multiuse spaces.

Vegetative Filter Strips

Filter strips are vegetated open spaces, such as meadows or forests, which treat stormwater by slowing it down and allowing sediment and associated pollutants to settle out of the runoff. The gently sloped vegetated area can also absorb stormwater as it spreads over the landscape, reducing the total runoff volume from small storm events (see Figure 5.24).

In order to prevent erosion, water is slowly dispersed across the filter strip in a shallow and relatively even sheet of water known as overland sheet flow. Gravel-filled trenches or rows of bunchgrass may be used along the leading edge of the strip as "level spreaders" to encourage sheet flow. Filter strips can receive water directly from impervious surfaces and are often part of a pretreatment practice for other stormwater management strategies. Using filter strips as part of a stormwater treatment train reduces sediment and particulates pollutant loads, thereby improving the effectiveness of the next treatment strategy.

■ FIGURE 5.24 Vegetative filter strip.

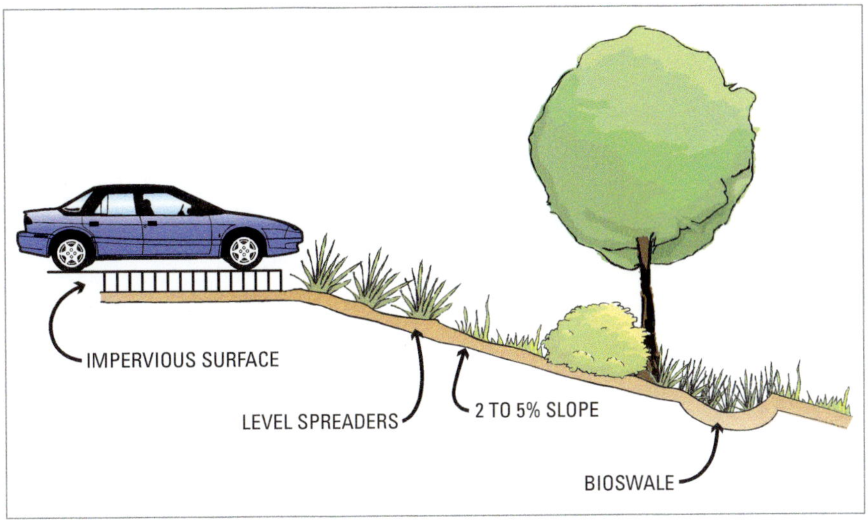

■ DESIGN CONSIDERATIONS

To reduce runoff and enhance water quality, filter strips require a gentle slope of 5 percent or less and dense stands of vegetation. Existing natural areas can be used as filter strips, or new landscapes can be constructed for the purpose. Where an existing natural area is used, the soil conditions and vegetation must be consistent with the stormwater management intent in order to avoid site damage. A small, permeable berm can be constructed at the downstream slope of the filter strip to temporarily pond water and further increase potential infiltration.

Soils should be porous and able to sustain healthy stands of vegetation without the use of fertilizers or other amendments that may contaminate the stormwater. The growth habitat and structure of vegetation affect its ability to successfully mange stormwater. For example, tall, deep-rooted grasses such as switchgrass (*Panicum virgatum*) are more effective at slowing runoff and removing pollutants than short turfgrasses (Smith 2000). Other advantageous vegetative characteristics include an upright structure that can withstand flooding and the ability to grow in both wet and dry soil conditions.

Bioswales

Bioswales, also known simply as swales, are vegetated channels that slowly convey, filter and infiltrate stormwater . The shallow channels can be planted with a variety of trees, shrubs, and herbaceous perennial plants, and designed to fit into almost any landscape setting (see Figure 5.25). When needed, drainage pipes and gravel reservoirs can be installed underneath bioswales to increase infiltration and storage capacity.

Similar to other biofiltration strategies, swales use soil, microorganisms, and vegetation to remove pollutants and clean the water. Bioswales have been shown to be very effective at removing sediment, oil, and grease, and, to a lesser extent, metals and nutrients (Jurries 2003). Vegetation slows the runoff and in doing so allows sediment and the attached pollutants to settle. Water carries pollutants into the soil, where they may be immobilized and/or decomposed by soil microorganisms and vegetation. The capacity of bioswales to remove pollutants can be improved by maximizing vegetative cover and the amount of water captured by a swale. This can be accomplished by planting a variety of plant types—trees, shrubs, grasses, and groundcovers. Tall, deep-rooted grasses have been found to be more effective at removing pollutants than short turfgrass. When selecting vegetation, caution should be taken to avoid overly restricting water flow (Jurries 2003).

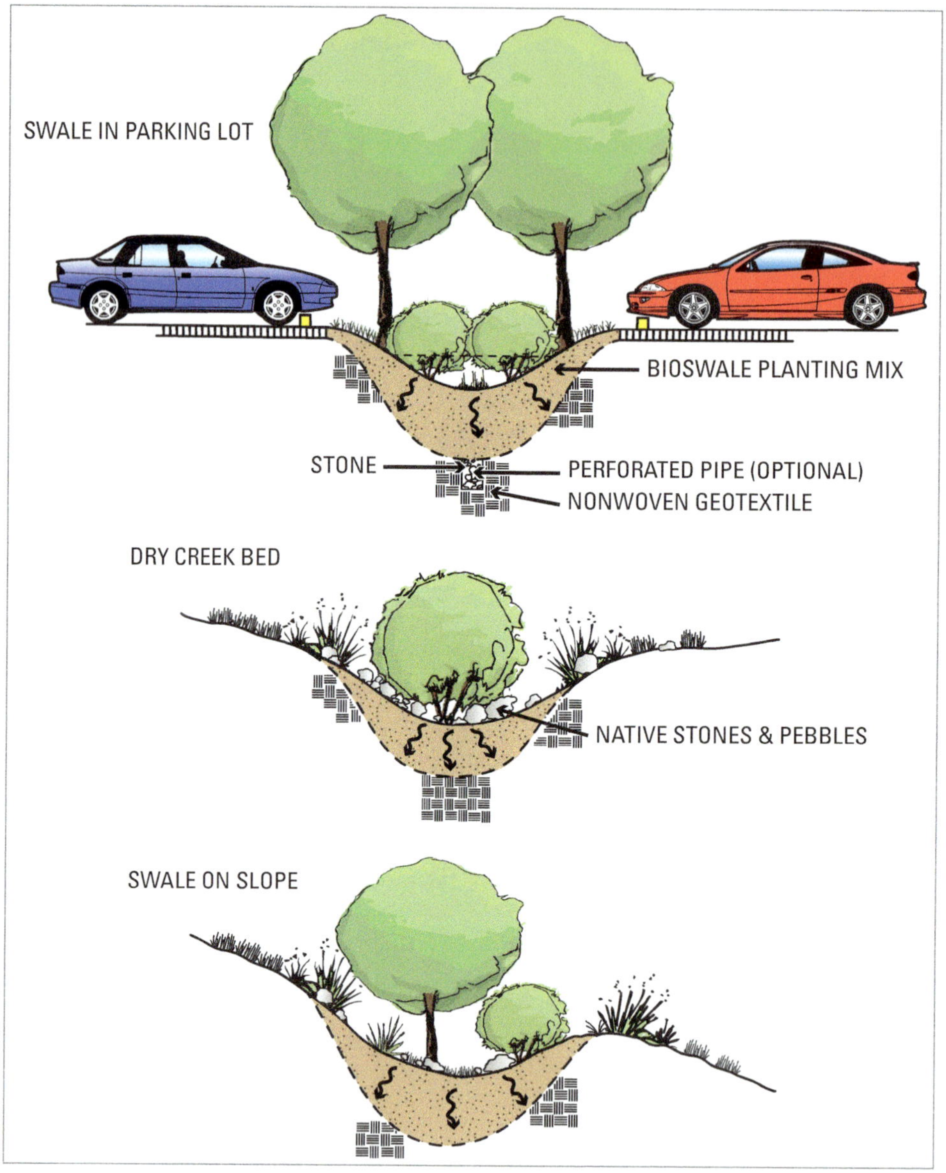

■ **FIGURE 5.25** Bioswale.

The cities of Portland, Oregon, and Seattle, Washington, have been very innovative in incorporating swales into existing neighborhood roadways, commercial parking lots, and other locations. Cost comparisons conducted by the Seattle Public Utilities found that the natural drainage systems deliver higher levels of environmental protection for receiving waters at a lower cost than traditional stormwater management improvements (City of Seattle 2007).

■ DESIGN CONSIDERATIONS

The slope of bioswales typically ranges from 1 to 6 percent, with the optimal slope being between 1 and 2 percent (Jurries 2003). Temporary water ponding can be encouraged through design features that slow the flow of water, such as tall, herbaceous vegetation or check dams. Soils underlying the swales are often amended with compost or other materials to increase their infiltration rate and storage capacity.

It is important to select vegetation that provides cover year-round and does not require additional irrigation or fertilizers, which can contribute to water pollution. By using both warm and cool season vegetation, biological activity in the soil can be better supported. The Oregon Department of Environmental Quality (Jurries 2003) outlines the following criteria that vegetation must meet in order to maintain channel stability and improve pollutant removal:

- Dense aboveground cover and root mass that can hold soil in place
- Upright structure that can be maintained during storm events to slow the runoff velocity and remove suspended pollutants
- Tolerate both periodic flooding and drought conditions
- Thrive within the growing conditions of the bioswale

It is important that swales do not become overly compacted and unable to absorb runoff or support healthy root development and soil biota. Construction and maintenance practices and any land uses that compact the soil should be avoided. Maintenance requirements are similar to a typical garden. Bioswales should be inspected annually and after major storm events to repair damage and ensure proper drainage.

RESOURCES

City of Portland Bureau of Planning and Sustainability: www.portlandonline.com

Dunnett, N., and A. Clayden. 2007. *Rain gardens: Managing water sustainably in the garden and designed landscape.* Portland, OR: Timber Press.

Jurries, D. 2003. Biofilters (bioswales, vegetative buffers, and constructed wetlands) for storm water discharge pollution removal. Portland, OR: State of Oregon Department of Environmental Quality.

Prince George's County, Maryland. 2007. *Biofiltration manual.* Prince George's County, MD: Department of Environmental Resources.

Rain Gardens

Rain gardens are shallow, vegetated depressions typically between 4 and 8 inches (10 and 20 cm) deep that collect and absorb runoff from the surrounding area (see Figure 5.26). To encourage quick drainage, the soil underlying the garden is highly porous and may be amended with compost or sand. Rain gardens are not intended to hold water for extended periods of time and are typically designed to draw down any pooled water within twenty-four to forty-eight hours. A gravel reservoir and perforated subdrainage pipe may be required in some circumstances to speed the drainage of the garden.

Rain gardens reduce stormwater volume by facilitating infiltration and evapotranspiration. A typical rain garden will infiltrate 30 percent more water than a conventional lawn (Dunnett and Clayden 2007). Pollutants carried by the runoff settle out and are filtrated as the stormwater moves through the surface mulch layer and amended soil.

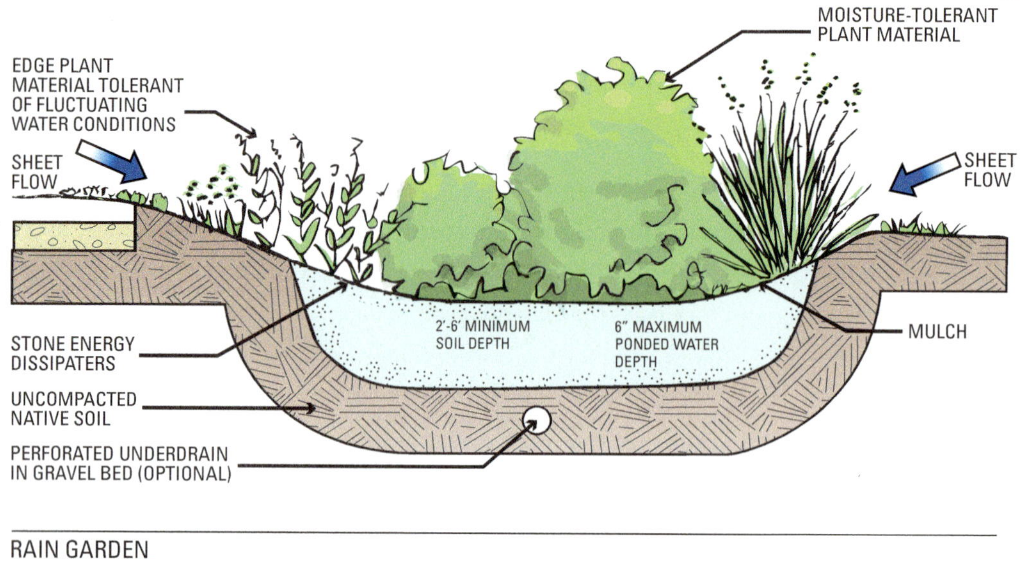

■ **FIGURE 5.26**
Rain garden.

■ DESIGN CONSIDERATIONS

Rain gardens should remain level in order to prevent water from overflowing before it has the chance to soak into the soil. In sloped areas, a pervious retaining wall on the down hillside can be added to maintain level soils. Rain gardens designed in this fashion are sometimes known as weep gardens (Prince George's County 2007).

Rain gardens placed in full sun have the greatest evapotranspiration rates and typically have a wider variety of plant options. Perennial plants that are native to the region, that do well in both temporarily wet and dry soil conditions, and that do not require fertilizers are commonly recommended. Because water does not pond in a rain garden for more than a few hours, the gardens have minimal safety and liability issues, and mosquitoes are typically not a problem (Prince George's County 2007).

Rain gardens require maintenance similar to that of a standard perennial garden. Typical maintenance practices include applying mulch, removing and replacing dead vegetation, and repairing eroded areas. The monitoring and maintenance plan should include indicators of success and failure, as well as strategies to ameliorate any issues. Possible indicators of failure include poor plant performance, extended water ponding, and putrid-smelling soils. Annual inspections are recommended after major storm events to repair damage and ensure there is proper drainage. Maintenance practices and land uses that compact the soil should be avoided.

■ RESOURCES

Dunnett, N., and A. Clayden. 2007. *Rain gardens: Managing water sustainably in the garden and designed landscape.* Portland, OR: Timber Press.

The Low Impact Development Center: www.lowimpactdevelopment.org

Prince George's County, Maryland. 2007. *Biofiltration manual.* Prince George's County, MD: Department of Environmental Resources.

■ FIGURE 5.27
Sunken infiltration planters and structural soil cleanse and infiltrate water at the Taylor 28 streetscape in Seattle.

Stormwater Planters

Stormwater planters are specialized containers designed to capture stormwater runoff and treat pollutants. The planting containers can be raised or located at ground level, and either freestanding or placed directly against a building as an extension of the outside walls and foundation. A variety of trees, shrubs, and perennials can be grown within each container.

Runoff enters the surface of the planters through roof downspouts, overland flow, or other plumbing. Stormwater planters typically are designed to drain pooled surface water within several hours. When runoff exceeds the infiltration capacity, excess water overflows to another stormwater treatment feature or is diverted to the conventional drainage system.

As with all stormwater management strategies that encourage infiltration, the planters reduce runoff volumes, which in turn decreases pollutant loads to receiving water bodies. Stormwater storage allows sediments and pollutants such as nutrients and metals to settle out of the water and be held or treated by the soil, microorganisms, and vegetation.

■ DESIGN CONSIDERATIONS

Stormwater planters are similar to rain gardens; however, due to greater depths the planters can often allow more water to be stored and treated. The planters can be used as design features that serve multiple purposes, such as screens, retaining walls, and bench seats. The shape and location of stormwater planters can be modified to fit almost any landscape, making the design strategy an optimal alternative for projects with limited space or other characteristics that may restrict the management of stormwater on-site.

There are two basic types of stormwater planters: infiltration and filtration. Infiltration planters have a series of pervious layers and direct contact with the soil below, allowing water to infiltrate into subsoil layers (see Figure 5.28). Due to changes in underlying soil moisture, the following criteria are commonly recommended for infiltration planters:

- Soils underlying the container should be porous.
- The planters should not be used within 10 feet (3 m) of most buildings or on slopes exceeding 10 percent.
- The bottom of the planter should be at least 24 inches (61 cm) above bedrock and a minimum of 36 inches (91 cm) above groundwater.

Filtration planters, also known as flow-through planters, are lined with an impervious layer to prevent water from infiltrating underlying soils (see Figure 5.29). The planters temporarily store and treat runoff but are not as effective at reducing runoff volume due to the impervious layers that prevent infiltration. Filtration planters are used where the underlying soil conditions prevent infiltration or where infiltration may create unsafe conditions such as building damage, high water tables, existing soil pollutants, or groundwater contamination.

■ **FIGURE 5.28**
Infiltration planter.

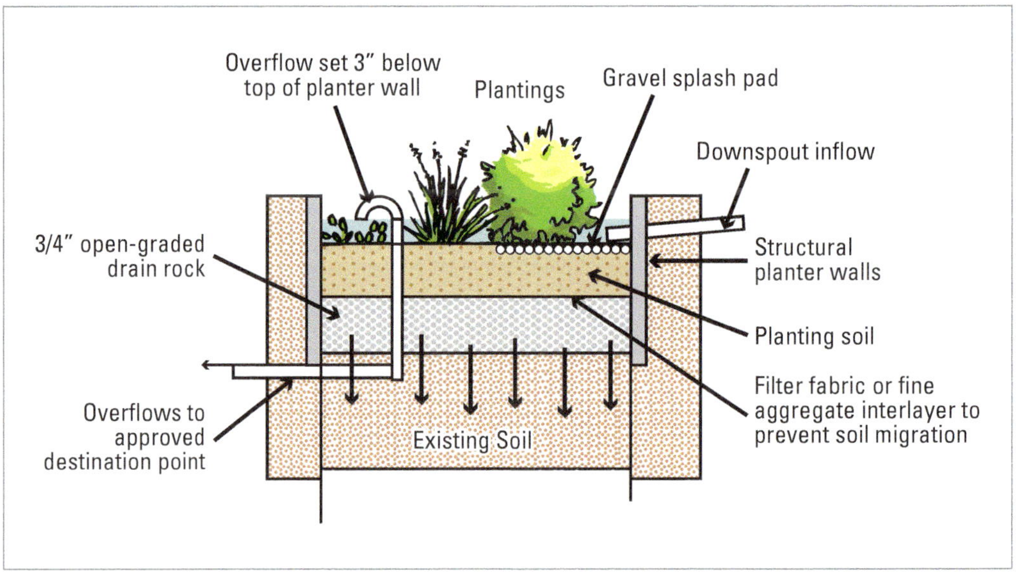

ADAPTED FROM PORTLAND STORMWATER MANAGEMENT SYSTEM

■ **FIGURE 5.29**
Filtration planter.

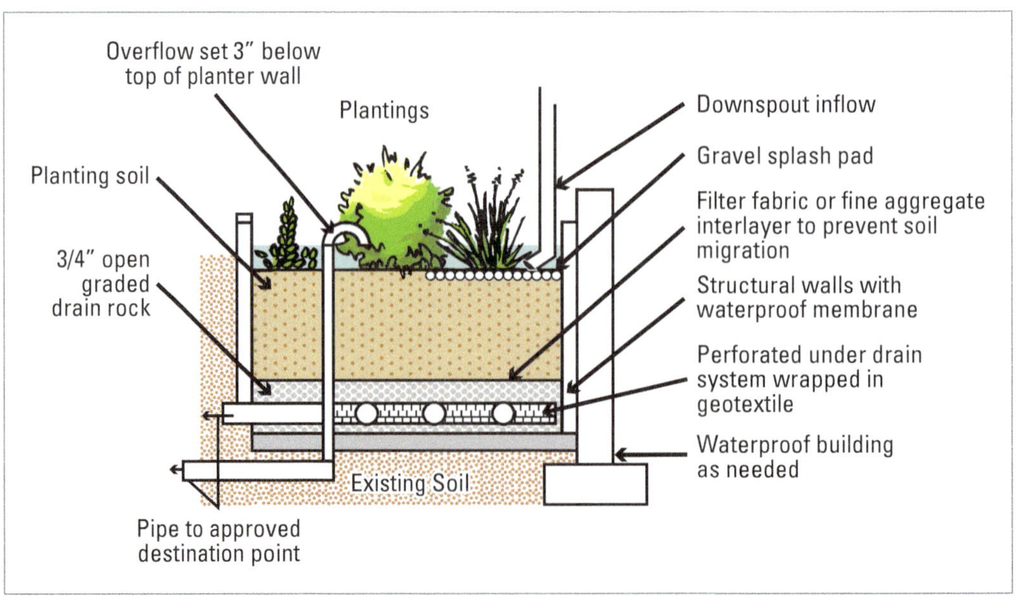

ADAPTED FROM PORTLAND STORMWATER MANAGEMENT SYSTEM

Plants that are chosen for the planters should be able to survive short periods of flooding and thrive without additional fertilizers or pesticides that may contaminate the water moving through the garden. Stormwater planters can be a seed source for downstream areas; therefore, invasive species or noxious weeds should be avoided and removed immediately.

The maintenance requirements of the planters are similar to other gardens; however, where there are large volumes of silt and clay being deposited in the planter, the excess material should be removed immediately to avoid clogging. As with all systems, the stormwater planters should be inspected annually and after major storm events to repair damage and ensure there is proper drainage.

■ RESOURCES

Center for Watershed Protection. 2010. *New York State: Stormwater Management Design Manual*. Albany, NY: Department of Environmental Conservation.

City of Portland, Oregon, Bureau of Environment Services. 2008. *Stormwater management manual*. Portland, OR: Bureau of Environment Services. www.portlandonline.com

Dunnett, N., and A. Clayden. 2007. Rain gardens: *Managing water sustainably in the garden and designed landscape. Portland, OR: Timber Press.*

Blue Roofs

Blue roofs are rooftop detention systems that moderate stormflows. Restriction devices located on the roof temporarily hold water back until it exceeds the established limits and overflows into the roof drain. Stormwater that is detained on the roof can be held until it evaporates or released to landscape areas. As a last resort, water can be slowly discharged to the sewer system after the storm surge has passed. The delayed release decreases pressure on the sewer system and the likelihood of combined sewer overflows.

■ DESIGN CONSIDERATIONS

Blue roofs require relatively flat, watertight roof surfaces that have the load-bearing capacity to support the additional weight. In many cases, only minor changes to the roof are required. Water on the roof can be concentrated in areas where it can be supported structurally and used to increase the evaporative cooling effect for the building (Foster et al. 2011). Blue roofs are less expensive to build and maintain than green roof systems; however, because vegetation is not present, they do not provide the same habitat, urban heat island mitigation, air quality, or aesthetic benefits.

■ RESOURCES

The City of New York. 2008. Sustainable Stormwater Management Plan 2008.

Foster, J., A. Lowe, and S. Winkelman. 2011. The *value of green infrastructure for urban climate adaptation*. Washington, DC: Center for Clean Air Policy.

REFERENCES

Bannerman, R. T., D. W. Owens, R. B. Dodds, and N. J. Hornewer. 1993. Sources of pollutants in Wisconsin stormwater. *Water Science and Technology* 28 (3–5): 241–59.

City of Portland, Oregon. Environmental Services. 2009. "How to manage stormwater: Downspout disconnection."

City of Portland, Oregon. 2010. "Downspout disconnection program." Retrieved November 17, 2010, from http://www.portlandonline.com/bes/index.cfm?c=43081&a=177712.

City of Seattle. 2007. *Seattle's natural drainage systems: A low-impact development approach to stormwater management.* Seattle Public Utilities, City of Seattle.

Daddow, R., and G. Warrington. 1983. *Growth-limiting soil bulk densities as influenced by soil texture.* Fort Collins, CO: USDA Forest Service, Watershed Systems Development Group.

Day, S. D., and S. B. Dickinson. 2008. Managing *stormwater for urban sustainability using trees and structural soils.* Blacksburg, VA: Virginia Polytechnic Institute and State University.

Department for Environment Food and Rural Affairs [DEFRA]. 2010. "Sewage overflows." Retrieved September 7, 2010, from http://www.defra.gov.uk/environment/quality/water/waterquality/sewage/overflows/.

Dunnett, N., and A. Clayden. 2007. *Rain gardens: Managing water sustainably in the garden and designed landscape.* Portland, OR: Timber Press.

Dunnett, N., and N. Kingsbury. 2004. *Planting green roofs and living walls.* Portland, OR: Timber Press.

Ferguson, B. K. 2005. *Porous pavements.* Boca Raton, FL: Taylor and Francis.

Haffner, T., and N. Bassuk. 2007. *Using porous asphalt and CU-Structural Soils.* Ithaca, NY: Urban Horticulture Institute, Cornell University.

Hunt, W., and E. Bean. 2006. North Carolina State University permeable pavement research and changes to the state of North Carolina runoff credit. Paper presented at the 8th International Conference on Concrete Block Paving, San Francisco.

Ingham, E., A. Moldenke, and C. Edwards. 2000. *Soil biology primer.* Ankeny, IA: Soil and Water Conservation Society in cooperation with USDA Natural Resources Conservation Service.

Jurries, D. 2003. Biofilters (bioswales, vegetative buffers, and constructed wetlands) for storm water discharge pollution removal. Portland, OR: State of Oregon Department of Environmental Quality.

Köhler, M., and M. Schmidt. 2003. Study of extensive "green roofs" in Berlin, part III: Retention of contaminants. Technical University of Berlin.

Loizeaux-Bennet, S. 1999. Stormwater and nonpoint-source runoff: A primer on stormwater management. *Erosion Control* 6 (7): 56–69.

Miller, G. T. 1998. *Sustaining the earth: An integrated approach.* Wadsworth.

National Oceanic and Atmospheric Administration [NOAA]. National Weather Service. 2010. "Hydrologic Information Center: Flood loss data." Retrieved October 3, 2011, from http://www.nws.noaa.gov/hic/flood_stats/Flood_loss_time_series.shtml.

Peper, P. J., E. G. McPherson, J. R. Simpson, S. L. Gardner, K. E. Vargas, and Q. Xiao. 2007. *New York City, New York municipal forest resource analysis.* Davis, CA: Center for Urban Forest Research, Pacific Southwest Research Station, USDA Forest Service.

Prince George's County, Maryland. 2007. *Biofiltration manual.* Prince George's County, MD: Department of Environmental Resources.

Schueler, T. R., and H. Holland. 2000. The importance of imperviousness. *Practice of Watershed Protection* 1 (3): 20-21.

Simmons, M., B. Gardiner, S. Windhager, and J. Tinsley. 2008. Green roofs are not created equal: The hydrologic and thermal performance of six different extensive green roofs and reflective and non-reflective roofs in a sub-tropical climate. *Urban Ecosystems* 11: 339–48.

Sites. 2009. Sustainable Sites Initiative: Guidelines and Performance Benchmarks 2009, Sites.

Smith, M. 2000. Vegetative filter strips for improved surface water quality. Ames, IA: Iowa State University Extension.

Thompson, J. W., and K. Sorvig. 2000. *Sustainable landscape construction: A guide to green building outdoors.* Washington, DC: Island Press.

Urban, J. 2008. *Up by roots: Healthy soils and trees in the built environment.* Champaign, IL: International Society of Arboriculture.

U.S. Composting Council [USCC]. 2008. USCC fact sheet: Compost and its benefits. Ronkonkoma, NY: USCC.

U.S. Environmental Protection Agency [EPA]. Office of Solid Waste and Emergency Response. 1997. Innovative uses of compost: Erosion control, turf remediation, and landscaping. Washington, DC: U.S. EPA.

_______. Office of Solid Waste and Emergency Response. 2001a. A citizen's guide to bioremediation. Washington, DC: U.S. EPA.

_______. Office of Solid Waste and Emergency Response. 2001b. A citizen's guide to phytoremediation. Washington, DC: U.S. EPA.

_______. Office of Water. 2001c. *Report to Congress: Implementation and enforcement of the combined sewer overflow control policy.* Washington, DC: U.S. EPA.

_______. Nonpoint Source Control Branch. 2007. *Reducing stormwater costs through low impact development (LID) strategies and practices.* Washington, DC: U.S. EPA.

_______. 2011. "Watershed Assessment, Tracking, and Environmental Results." Retrieved April 11, 2011, from http://iaspub.epa.gov/waters10/attains_nation_cy.control#total_assessed_waters.

Xiao, Q., and E. G. McPherson. 2002. Rainfall interception by Santa Monica's municipal urban forest. *Urban Ecosystems* 6:291–302.

Xiao, Q., E. G. McPherson, S. L. Ustin, and M. E. Grismer. 2000. A new approach to modeling tree rainfall interception. *Journal of Geophysical Research* 105 (D23): 29, 173–88.

BIGSTOCK/WARREN PRICE

■ **FIGURE 6.1**
Receding shorelines at Lake Buchanan, the first of a series of reservoirs along the Colorado River, are the result of prolonged drought in Central Texas.

CHAPTER 6

Sustainable Solutions: Water Shortages

GLOBAL FRESHWATER SHORTAGES are quickly becoming one of the most urgent challenges facing humanity. Scientists estimate that by 2030, half of the world's population will have inadequate supplies of freshwater (WWAP 2009). In the United States, thirty-six of the fifty states anticipate freshwater shortages in the next ten years (U.S. GAO 2003). One-fifth of the European population, that is, approximately 113 million people, currently live in water-stressed regions (EEA 2010). In India and parts of northern China, groundwater tables are falling at a rate of more than 3.3 feet (1 meter) per year (Watkins 2006). Brownouts are occurring in Brazil and South Africa because there is not enough water to drive hydroelectric power plants. And, over the last decade, Australia has experienced severe drought conditions, leaving reservoirs at record lows and prompting extensive water restrictions.

As the burdens of freshwater shortages become more pronounced, all water usage will be questioned and additional conservation measures will be required. Historically, water conservation has been viewed as a standby or temporary measure, emphasized during periods of drought; however, chronic water shortages resulting from urban population growth and wasteful water usage are making the permanent reduction of water use standard practice for many communities (Asano et al. 2006). Sustainable sites can advance conservation efforts by creating drought-resistant landscapes that do not rely on potable water but reuse on-site nonpotable water resources such as stormwater runoff, air-conditioner condensate, harvested rainwater, and greywater. Additional water requirements can also be minimized through the restoration and long-term management of soil health and the careful selection and maintenance of vegetation that is well suited to site conditions.

This chapter explores site strategies for reducing water waste and recharging groundwater supplies. Alternative on-site water resources that can replace potable water in the landscape are discussed, as are microdetention strategies that slow and capture water on-site for reuse, and the importance of soil stewardship and appropriate plant selection.

■ GLOBAL POPULATION AND THE WORLD'S WATER SUPPLY

Over the next forty years, the global population is expected to increase from 6 billion to an estimated 9 billion, yet the world's water supply is constant. Only 3 percent of the global water supply is fresh; the majority of it is locked in ice or stored deep in the earth, making its extraction very expensive. The remaining 97 percent is found in the oceans and is too salty for human consumption, irrigation, and industrial uses. Water from the oceans can be processed; however, desalination is an energy-intensive practice. In addition, the concentrated brine discharge, the by-product of desalination, contains large amounts of salt and other minerals that are damaging to terrestrial and aquatic ecosystems and can be difficult to dispose of safely.

Freshwater Shortages: The Cause

Water is a renewable but limited resource. Shortages result when consumption outpaces its resupply. All too often, the problem is not an extreme shortage of water but an unwillingness to live within the local water budget. Many regions mine or transport copious amounts of water to support development that cannot be sustained over the long term. This type of activity has the unfortunate consequence of making cities and industries dependent on political agreements and technological advances to deliver fresh water.

People, particularly those in wealthy nations, waste a considerable amount of water on a daily basis. This is largely due to easy access and artificially low water prices that do not reflect the true value of the resource. Low water prices undermine incentives for water conservation by unwittingly conveying the message that water is plentiful and will always be readily available.

A significant portion of the water consumed in urban environments is used for landscape irrigation. On average, 30 percent of the water consumed in U.S. households is devoted to outdoor uses, such as watering lawns and gardens (U.S. EPA 2007). In drier regions, such as the American Southwest, irrigated landscapes can account for 45 to 70 percent of total residential water consumption (ADRW 2011).

Urban environments also contribute to water shortages by converting vegetated land into impervious surfaces—such as roads, roofs, and parking lots—that prevent groundwater recharge. In addition, conventional stormwater management practices quickly transport runoff from developed landscapes to off-site locations before it has had an opportunity to benefit the site. As a result, soil moisture and groundwater supplies are greatly reduced.

Water Shortages: How They Affect Our Lives

Water supports our environment and sustains our lives. It is essential to human survival, our livelihoods, and almost every form of economic production. This finite resource is required for producing food, clothing, and electronics; transporting our waste; and supporting the natural environment (see Figure 6.2).

The effects of water shortage can occur quickly and have lingering repercussions. For example, during a four-month period in the summer of 1988, the central and eastern United States suffered over $40 billion in economic damages due to water shortages. In Texas, the Water Development Board has determined that if the state does not ensure adequate water supply for future generations, it will have 7.4 million fewer jobs and 38 percent less income by 2050 (U.S. GAO 2003).

Burgeoning populations and growing water demands are fueling local and international conflicts as individuals, cities, states, and nations fight for the resource integral to their existence. One example is the Mekong River in Asia, which winds through six countries and supplies water to an estimated 65 million people. China's damming of the Mekong for irrigation and energy production has produced tensions with downstream users in Laos, Vietnam, and Thailand, where water shortages have reduced fish populations and food production (Fuller 2010). In the United States, water wars are brewing between seven states in the arid Southwest over control of the Colorado River, which is the water source for 30 million people and 3 million acres of farmland (Matalon 2010).

Competition for water not only impacts economic production but also the viability of many ecosystems. The overconsumption of water in urban areas often leads to reduced water availability for native ecosystems, resulting in environmental degradation, loss of habitat, and reduced recreational opportunities.

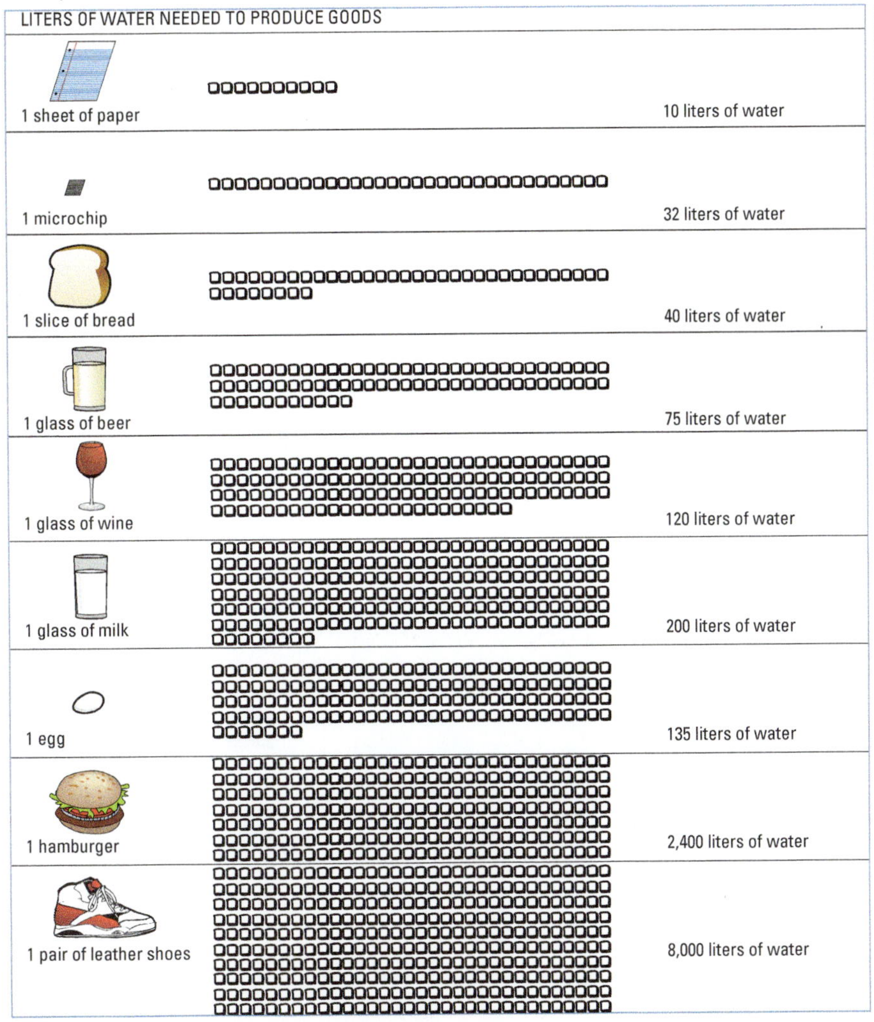

DATA SOURCE: HOEKSTRA AND CHAPAGAIN 2006

■ **FIGURE 6.2** Water needed to produce goods.

Sustainable Site Strategies to Alleviate Water Shortages

Water is a limited resource that society no longer has the luxury of wasting. Landscape practitioners can lead conservation efforts and sustain water supplies for future generations by pursuing the goal of no potable water use in the landscape. Projects are more likely to reach this goal if it is established early in the design process and supported by all team members. It can be accomplished through soil stewardship practices that protect and restore soil health, the careful selection and maintenance of vegetation that is well suited to site conditions and stringent water conservation. Potable water use in the landscape can also be safely replaced with alternative on-site water resources that are often wasted, such as stormwater runoff, harvested rainwater, air-conditioner condensate, and greywater (see Figure 6.3).

A water-balance analysis, or water budget, that estimates a site's water requirements, as well as the amount of water available from precipitation and alternative on-site resources, should be conducted early in the project and used to guide design decisions. Such an analysis is useful when determining the size of irrigated areas, appropriate plant types, and water catchment and storage requirements, and is critical for ultimately maintaining water demands within the on-site water availability. Design decisions should be constantly weighed throughout the development of the project and adjustments made to balance the site's water use with the available nonpotable water resources (see Figure 6.4).

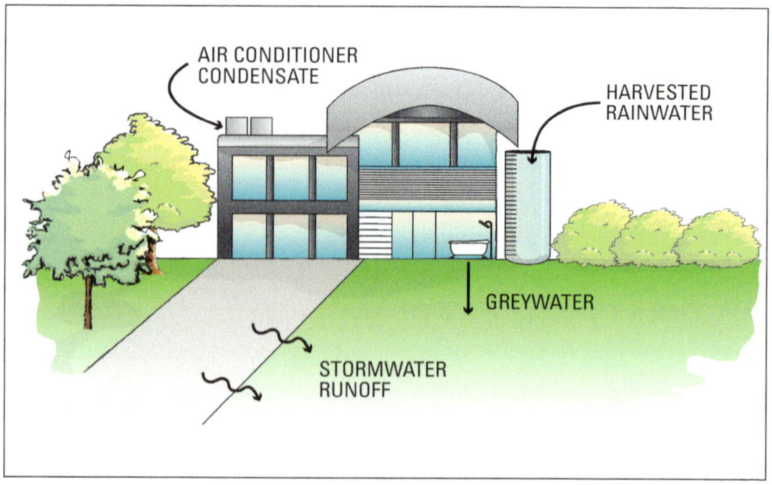

HEATHER VENHAUS

■ **FIGURE 6.3**
Alternative on-site water resources that can safely replace potable water.

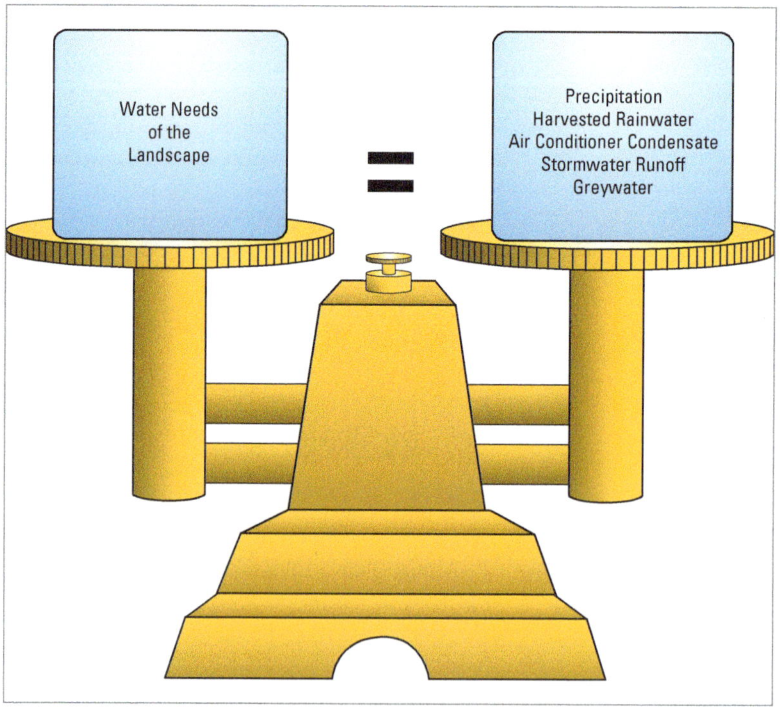

HEATHER VENHAUS

■ **FIGURE 6.4**
Water balance occurs when the water requirements of a landscape equal the on-site water resources from precipitation, harvested rainwater, air conditioner condensate and greywater.

Site strategies to minimize potable water and recharge groundwater supplies include:

- Stormwater catchment and reuse
- Greywater catchment and reuse
- Reclaimed water reuse
- Air-conditioner condensate catchment and reuse
- Drought-resistant soils and vegetation
- Avoiding wasteful irrigation and maintenance practices

It is easy to overlook the need for judicious water use and monitoring when alternative water sources are being used. Oftentimes alternative water sources are treated as "extra" water available for noncritical functions. Project teams and maintenance staff must understand that alternative water sources are as valuable as potable water and should be treated as such. Site monitoring and maintenance plans should include guidance on conservation strategies as well as on the proper monitoring and care of the alternative water systems.

TABLE 6.1
Alternative Water Sources for Sites

STRATEGY	DESCRIPTION	POTENTIAL QUANTITY OF WATER AVAILABLE	TYPICAL USES
Downspout disconnection	Separate roof downspouts from sewer systems and direct runoff to landscape.	Approximately 600 gallons (2,271 L) of water can be collected for each inch (2.54 cm) of rain falling on a 1,000-foot^2 (93 m^2) catchment surface. Downspouts typically receive water from specific sections of the roof, not the entire roof area.	Overland irrigation and groundwater recharge.
Passive rainwater harvesting	Divert rainwater overland from impervious surfaces to vegetated areas for immediate use.	Approximately 500 gallons (1893 L) of water can be collected for each inch (2.54 cm) of rain falling on a 1,000 ft^2 (93 m^2) ground level impervious catchment surface.	Overland irrigation and groundwater recharge.
Active rainwater harvesting	Capture rainwater in a cistern or tank for reuse at a later date.	Approximately 600 gallons (2,271 L) of water can be collected for each inch (2.54 cm) of rain falling on a 1,000-foot^2 (93 m^2) catchment surface.	Irrigation, toilet flushing, groundwater recharge, and makeup water for cooling equipment. Rainwater can be treated to become potable and used for drinking and other domestic purposes.
Greywater reuse	Wastewater from clothes washers, showers, bathtubs, and lavatory faucets. Greywater does not include toilet water, known as sewer or blackwater.	A typical U.S. household generates an average of 35 gallons (132 L) per person per day (Roesner et al. 2006). Other building types may vary.	Subsurface irrigation, groundwater recharge, and toilet flushing.
Reclaimed water	Treated and purified outflow from municipal wastewater treatment plant.	Water volumes can be large and are dependent upon contract with the local water authority.	Irrigation, fire protection, groundwater recharge, toilet flushing, ornamental landscape features not intended for human contact, surface-water augmentation, industrial cooling, and process water.
Air-conditioner condensate	Natural by-product of air-conditioning systems. Condensation occurs when water vapor in the indoor air—often described as humidity—comes in contact with the cooling components of air conditioning equipment.	The amount of condensate generated is dependant upon the local climate, building use, and air-conditioning system. An estimated 3 to 10 gallons of condensate is generated per day per 1,000 feet^2 of air-conditioned space (11.35 to 37.84 L /day/92.9 m^2) (Alliance for Water Efficiency 2010).	Irrigation, toilet flushing, ornamental landscape features not intended for human contact, groundwater recharge, makeup water for cooling equipment.

■ CASE STUDY

TAYLOR 28

PROJECT TYPE: Streetscape

LOCATION: Seattle, Washington

SIZE: 38 square feet (11.6 m^2) wide, 391-foot-long streetscape, including 15,000-square-foot (4,572 m^2) pedestrian plaza

COMPLETION DATE: 2009

HIGHLIGHTED SUSTAINABLE PRACTICES:

- Urban infill redevelopment
- Reallocation of underutilized roadway for pedestrian access
- Zero potable water use for all on-site and right-of-way landscape irrigation
- Zero discharge for a twenty-five-year storm event, which greatly reduces drainage to Seattle's overburdened combined sewer system

THE SITE: Downtown infill site near the Space Needle in the Denny Triangle neighborhood. Formerly a wide asphalt street with 45-degree parking on either side.

■ FIGURE 6.5
Taylor 28 is designed as a complete street that supports bikes, pedestrians, placemaking, and green infrastructure systems.

Design Overview

Taylor 28 is the first residential, mixed-use development within the transforming Denny Triangle neighborhood. The project sets a precedent for a new urban design standard that transfers underutilized roadway into the public realm (see Figure 6.5). The new pedestrian-focused neighborhood enhances the quality of the urban experience and contributes to a healthier Puget Sound by minimizing stormwater runoff and input to Seattle's overburdened combined sewer overflow pipe system.

Prior to development, Taylor Avenue included two travel lanes and back-in angled parking on both sides of the street. The project's design maintains the same vehicular volume (two travel lanes) but eliminates the inefficient angled parking. The final design resulted in a reduction in vehicular width of 20 feet (6 m) while still maintaining some parallel on-street parking. This design concept has been approved by the city of Seattle for the entirety of Taylor Avenue, which stretches several blocks north and south of the project site.

The project is designed to manage stormwater up to a twenty-five-year storm event. Stormwater management strategies, such as permeable concrete, infiltration planters, and on-site rainwater harvesting, achieve zero discharge for both on-site and right-of-way runoff at the sidewalk level. The majority of the stormwater is managed with a 16,000-gallon (60,567 L) rainwater cistern that provides water reuse for nonresidential toilet flushing and is also the sole water

TAYLOR 28 *(CONTINUED)*

source for all on-site and right-of-way landscape irrigation. In the winter, when irrigation is not necessary, the cistern supplies water for toilet flushing within the building. The dual-use cistern allows the site to maintain a balance between an adequate water supply and available cistern capacity. This rainwater reuse strategy, in addition to efficient low-flow fixtures, saves up to 122,000 gallons (461,820 L) of potable water per year (see Figure 6.6).

Taylor 28 was designed to catalyze the neighborhood development by creating great public space and attracting more residents through a combination of apartments and retail. The design team worked closely with key city of Seattle staff to achieve outcomes that crossed typical boundaries between zoning, planning, streets, and utilities to address layout, maintenance responsibilities, rainwater harvesting and reuse, and stormwater collection and distribution.

■ **FIGURE 6.6**
Rainwater is captured in a 16,000-gallon (60,567 L) dual-use rainwater cistern that provides water reuse for nonresidential toilet flushing and is also the sole water source for all on-site and right-of-way landscape irrigation. Water from the plaza flows into vegetated infiltration rain gardens, where it is cleansed and infiltrated back into the water table.

PROJECT TEAM

- **MITHUN:**
  www.mithun.com
  Jim Bodoia, Architecture
  T. Frick McNamara, Landscape Architecture
  Max Anderson, Architecture
  Mat Lipps, Architecture
  Dave Pawlowski, Architecture
  Sara Raab, Landscape Architecture
  Lauren Acheson, Landscape Architecture
  Chuck Weldy, Construction Administration
  Suzan Schneider, Project Administration
- **BUSH, ROED & HITCHINGS, INC., CIVIL ENGINEERING**
  www.brhinc.com
- **GOLDER ASSOCIATES, GEOTECHNICAL ENGINEERING**
  www.golder.com
- **HEFFRON TRANSPORTATION, INC., TRAFFIC ENGINEERING**
  www.hefftrans.com
- **INTERFACE ENGINEERING, ELECTRICAL AND MECHANICAL ENGINEERING**
  www.interfaceengineering.com
- **YU & TROCHALAKIS, PLLC, STRUCTURAL ENGINEERING**
  www.ytengineers.com
- **FAULKNER DESIGN GROUP, INTERIOR DESIGN**
  www.faulknerdesign.com
- **SCOTT AG, SIGNAGE DESIGN**
  www.scottag.com

Stormwater Catchment and Reuse

Stormwater runoff is rain or snowmelt that accumulates and flows overland instead of soaking into the ground. Conventional stormwater management practices typically channel stormwater quickly and efficiently away from developed areas into detention ponds, receiving water bodies, or storm sewer systems. The unintended consequence of conventional stormwater management is an increase in both the number and severity of flood events, water pollution, and the disposal of a valuable water resource.

Sustainable sites retain and reuse stormwater to the greatest extent possible. This is achieved through various microdetention strategies that slow and capture water on-site, where it is reused in the landscape and buildings.

Strategies for accomplishing stormwater catchment and reuse include:

- Rainwater harvesting
- Downspout disconnection, see Chapter 5
- Rain gardens, see Chapter 5
- Bioswales, see Chapter 5
- Stormwater planters, see Chapter 5

RAINWATER HARVESTING

Prior to the proliferation of centralized water systems and affordable wells, rainwater was commonly collected in many parts of the world for household, landscape, and agricultural purposes. Water shortages, health concerns, and economic incentives have renewed interest in this time-honored and relatively simple practice.

Rainwater can be harvested from any surface, such as roofs, roads, driveways, and parking lots, that can capture and convey water. The water can be temporarily stored or used immediately to support the needs of landscapes or buildings. Rainwater harvesting has many benefits, including increased water availability, reduced stormwater runoff, and pollutant capture. It can also serve as an education tool that adds interest to a site and makes the movement and storage of water obvious and artful.

■ PASSIVE RAINWATER HARVESTING

Diverting rainwater overland to vegetated areas for immediate use is referred to as passive rainwater harvesting. Impervious surfaces such as roads, driveways, parking lots, and sidewalks can be designed to direct runoff to landscape areas instead of storm drains. Receiving areas can be linked so that overflow from one microbasin naturally drains into another. This biofiltration practice improves the water-cleansing potential of the landscape and significantly reduces or eliminates runoff (see Figures 6.7 and 6.8).

SOURCE: © LANCASTER 2004; ARTIST: JOE MARSHALL; REPRODUCED BY PERMISSION

■ FIGURE 6.7
Street-side microbasins. Curb cuts allow stormwater runoff from the street to flow into the microbasins, where it irrigates the landscape and reduces stormwater runoff. Surplus water flows from one basin to the next.

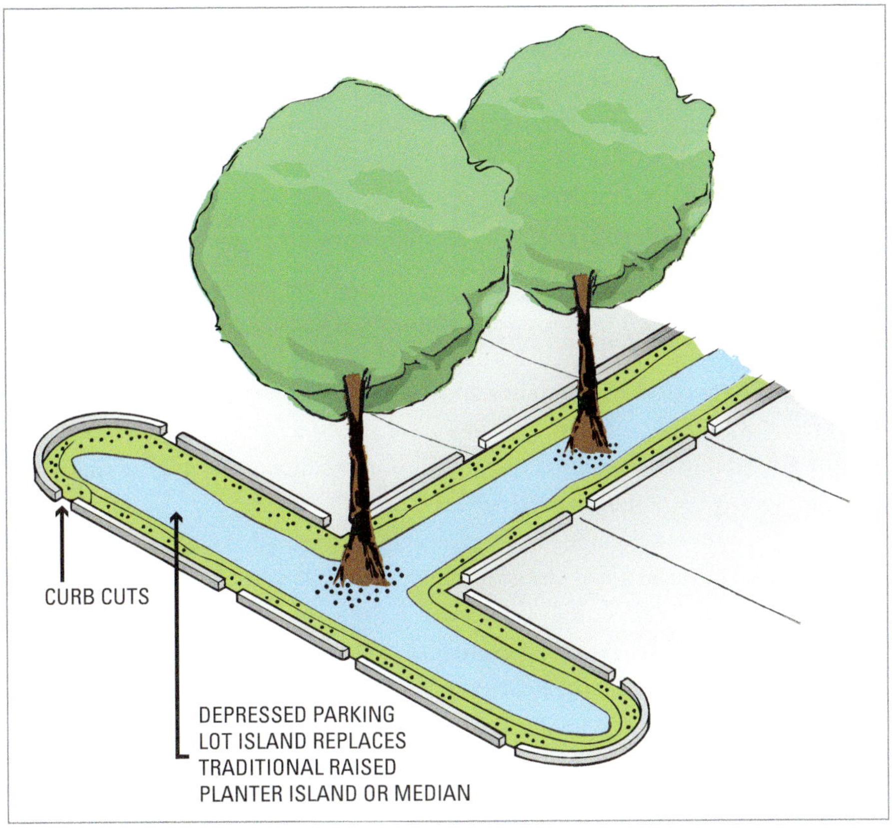

■ **FIGURE 6.8**
Parking lot island swale. Depressed parking lot captures runoff from the surrounding area.

■ ACTIVE RAINWATER HARVESTING

Active rainwater harvesting captures and stores rainwater for reuse. Water can be collected from rooftops, driveways, or other impervious surfaces and stored in a variety of tanks or cisterns (see Figure 6.9). Water entering storage containers is initially filtered to remove coarse debris, and can be used to irrigate the landscape, or in buildings for nonpotable water needs, such as toilet flushing and washing machine use. Rainwater can also be put through additional treatment processes and used for other household purposes such as bathing, cooking, and drinking.

■ **FIGURE 6.9**
An 8,000-gallon (30,282 L) rainwater harvesting cistern greets visitors at the Lady Bird Johnson Wildflower Center. Rainwater runoff from the roof of the auditorium travels down aqueducts into the cistern, where it is used to irrigate surrounding gardens.

Rainwater cisterns and tanks are manufactured in a variety of shapes and sizes and are commonly made from fiberglass, polyethylene, or galvanized metal. The tanks may be buried belowground or integrated into the landscape or building design as a freestanding structure or architectural element. Bladder systems that swell when full, then collapse as the water drains, are also available. These flexible systems can be easier to install than rigid cisterns and may be appealing for sites with open areas under existing buildings or decks.

Active rainwater harvesting is prohibited in some areas. Local water laws should be understood prior to designing a rainwater collection system.

DESIGN CONSIDERATIONS

Passive rainwater harvesting can be conducted through a variety of practices that slow runoff and encourage infiltration, such as bioswales, filter strips, terraces, and rain gardens. The plant palette and soil conditions must be carefully designed and maintained to accommodate the additional rainwater without eroding the soil or damaging vegetation.

When selecting catchment surface materials, the intended end use of the harvested rainwater must be considered. Surfaces along the ground plain, as well as roofing materials such as asphalt, asbestos, chemically treated wood shingles, and some painted roofs can release toxic materials into rainwater and should be limited to nonpotable purposes (Texas Water Development Board 2005). It is recommended that runoff from the intended catchment surfaces be tested prior to determining the treatment method and potable or nonpotable uses.

The amount of rainwater available for reuse depends on the size of the catchment and storage area, the efficiency of the collection system, and the amount of rainfall. Project teams should consider the median monthly rainfall when determining water availability. For estimation purposes, approximately 600 gallons (2,271 L) of water can be collected for each inch (2.54 cm) of rain falling on a 1,000-square-foot (93 m^2) impervious catchment surface. Surface materials impact the amount of runoff due to evaporation and minor infiltration; porous and rough surfaces are more likely to retain water and reduce runoff. Air-conditioning condensate can be collected and combined with harvested rainwater to increase the available water supply.

Almost any watertight structure or container can be used as a cistern; however, all tanks should be nontoxic and have a tight-fitting lid to prevent issues with mosquitoes or other pests. The Food and Drug Administration or equivalent agency should approve all cisterns intended for potable water use (Texas Water Development Board 2005).

The size, shape, and location of tanks impact water temperature and freezing potential. The ideal location for a cistern is between the rainwater source and the area of reuse. Cisterns can be placed aboveground or belowground and can be connected to the site's plumbing and irrigation systems. A benefit of aboveground tanks is that the head pressure may be enough to transport water to other areas of the site without requiring additional energy for pumping. Aboveground tanks can also be incorporated into the overall design as an interesting and educational amenity. Belowground systems can moderate water temperatures, reduce algal and bacterial growth, and save space by being placed under existing structures or paved areas.

The site-monitoring and maintenance plan should include descriptions and details of the rainwater harvesting water system, along with troubleshooting guidance on how to identify malfunctions. Monitoring the amount of water used from the system is a useful strategy for avoiding waste and identifying any leaks or malfunctions.

■ RESOURCES

Hopper, L., ed. 2007. *Landscape architectural graphic standards: Student edition.* Hoboken, NJ: John Wiley & Sons.

Kinkade-Levario, H. 2007. Design for water: Rainwater harvesting, stormwater catchment, and alternative water reuse. Gabriola Island, Canada: New Society.

Texas Water Development Board. 2005. *The Texas manual on rainwater harvesting.* Austin. www.twdb.state.tx.us

Greywater Catchment and Reuse

Greywater is wastewater from clothes washers, showers, bathtubs, and lavatory faucets that can be safely reused on-site for landscape irrigation, constructed wetlands, and toilet flushing (see Figure 6.10). Greywater does not include toilet water, which is known as sewer water or blackwater, and in some locations also excludes water from kitchen sinks and dishwashers due to their high levels of organic matter, oils, and grease. Approximately 50 to 80 percent of residential wastewater is comprised of greywater (Ludwig 2009). A typical U.S. household generates an average of 35 gallons (132.5 L) of greywater per person per day (Roesner et al. 2006). Greywater has been reused informally around the world for centuries; as water shortages continue to increase, the practice is gaining acceptance and in some cases is being promoted by environmental organizations and government agencies.

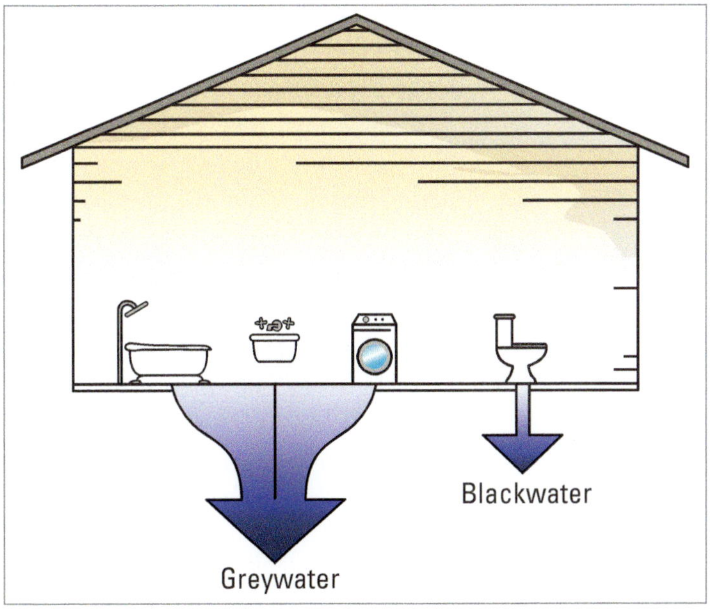

■ **FIGURE 6.10** Greywater sources. The typical U.S. household generates 35 gallons per person per day.

Using greywater to irrigate the landscape has a wide range of benefits, including:

- Providing a steady water source year-round and during times of drought
- Saving freshwater and potable water resources
- Decreasing the demand on water treatment plants and sewer systems
- Reducing energy use and the generation of greenhouse gases
- Recharging groundwater
- Improving awareness of and sensitivity to water use
- Reducing water bills

There are a variety of greywater systems, and they differ in their efficiency, complexity, and cost. Greywater systems often rely on gravity for water transport and use ordinary plumbing components. Dual plumbing is required to keep greywater separate from blackwater. Systems can be designed to convey greywater directly from the source to the landscape or can be plumbed to capture, treat, and temporarily store it for later reuse. Greywater often contains organic matter such as hair, skins cells, and clothing fibers, which are broken down by bacteria in the water, causing anaerobic and putrid water conditions. For this reason, greywater does not store well, and most systems quickly use the water for irrigation purposes. Filters remove hair, lint, and other large particles to avoid clogging irrigation systems. Due to health concerns, direct contact with greywater is not recommended, and immediate water-to-soil contact is often required. Aboveground sprinkler systems are typically prohibited due to the potential for direct human contact with the untreated water.

Art Ludwig, an ecological systems designer with over twenty-five years' experience designing wastewater systems, and author of *Create an Oasis with Greywater: Choosing, Building and Using Greywater Systems,* has observed that the overwhelming majority of greywater systems that remain in operation after ten years are very simple systems requiring few materials to construct and no energy to operate. However, more complex systems may be required by local governments and can offer greater efficiency, convenience, and a higher degree of treatment (Ludwig 2009). Common greywater treatments include particle removal, disinfection, and biological treatment (Asano et al. 2006).

DESIGN CONSIDERATIONS

In order to optimize the benefits and reduce the cost of greywater systems, an integrated design approach that includes both building and landscape professionals must be taken. Exploring options for greywater reuse early in the design process will improve the likelihood of reuse and the development of a cost-effective and efficient system.

Greywater laws and policies vary by region. The design team should clarify the local code's definition of greywater and identify any limitations or restrictions that may apply. Greywater reuse may not be economically feasible when codes require complex systems or when water availability is low. In very cold climates, freezing can prevent the use of greywater for portions of the year and may require special cold-weather adaptations such as subsurface piping, insulation, or the ability to temporarily drain to the sewer system.

Greywater retrofits on existing buildings require easy access to the wastewater plumbing. When this is not an option, greywater reuse is typically not cost-effective. New construction provides the opportunity to plumb the building for reuse from the start, saving time and money. Greywater pipes are typically identified by the color purple.

In some areas, greywater can also be used to flush toilets. Water-quality requirements for toilet flushing vary among regions, and due to human health concerns, the disinfection of greywater, which necessitates more sophisticated systems, is commonly recommended. Because of this requirement, it is typically more economically feasible to reuse greywater on-site for irrigation purposes.

The main risks associated with greywater reuse arise from physical contact and ingesting foods that have been contaminated by greywater. Research investigating the level of human health risks associated with greywater reuse have produced conflicting results; however, any risk to people can be minimized by using collection and distribution systems that avoid human contact and do not require user intervention.

A site's soil and geologic conditions must be understood in order to avoid groundwater contamination or water ponding that can lead to human contact or runoff. Site conditions that should be explored include:

- Soil texture and permeability
- Soil depth
- The presence of porous or fractured geologic features that would enable greywater to bypass soil purification and flow directly into groundwater (Ludwig 2009)
- The potential for flooding
- The area available for drainage
- Landscape topography and potential for runoff

The quality of greywater varies and depends on the source. Greywater typically contains salts, nutrients, and other organic compounds that have been washed from our skin or added to the water via soaps, detergents, or other chemicals. Sodium, potassium, and calcium salts often found in greywater can increase the alkalinity of the soil. High salinity can damage soil structure and limit the ability of roots to absorb water. Applying greywater over a broad area will help to avoid the buildup of harmful substances. In addition, high concentrations of salts and other water-soluble chemicals and nutrients can be flushed from the soil by rain and freshwater irrigation. Vegetation can also be used to bioremediate salts and other pollutants. The organic matter and nutrients found in greywater can, however, be beneficial to plants and soil microorganisms (see Figure 6.11) (Roesner et al. 2006). Both the potential risks and benefits should be weighed prior to the incorporation of a greywater system.

HEATHER VENHAUS

■ **FIGURE 6.11** Cenizo (*Leucophyllum frutescens*) (a) and fig (*Ficus spp.*) (b) are examples of vegetation that has been successfully grown in gardens irrigated with greywater.

Not all vegetation grows well when irrigated with greywater. Plants that typically grow in acidic soils, such as rhododendron or gardenia, tend to have difficulty (Ludwig 2009) due to the high pH of greywater. Plants that grow well in alkaline soils are commonly recommended, and vegetation irrigated with greywater should not be overly sensitive to elevated concentrations of salts. Generally, seedlings and young plants tend to be more sensitive to elevated salinity than well-established vegetation. Examples of salt-tolerant vegetation include bouganvillea (*Bouganvillea spectabilis*), cenizo (*Leucophyllum frutescens*), and Italian stone pine (*Pinus pinea*).

In order to decrease negative impacts on the soil and vegetation, the site-monitoring and maintenance plan should educate the site's caretakers about the following requirements of the greywater system:

- Local code requirements
- Description and details of the greywater system
- Instruction on how to properly operate the greywater system
- Monitoring and maintenance schedule—the greywater system should be checked regularly to ensure it is functioning properly and for issues such as leaks, breaks, or clogs
- Landscape-friendly detergents and cleansing agents
- Health and safety measures such as secure storage tanks, mosquito prevention, and avoiding the application of greywater to saturated soils
- Guidance on how to identify and mitigate damage to vegetation or soil that may be caused by the use of greywater

RESOURCES

Asano, T., F. Burton, and H. Leverenz. 2006. *Water reuse: issues, technologies, and applications.* New York: McGraw-Hill.

Ludwig, A. 2009. Create an oasis with greywater: Choosing, building, and using greywater systems. Santa Barbara, CA: Oasis Design.

Water Environment Research Foundation: www.werf.org

CASE STUDY

CATALINA FOOTHILLS

PROJECT TYPE: Singe-family residential

LOCATION: Tucson, Arizona

SIZE: 3 acres (1.2 hectares)

COMPLETION DATE: 2007

HIGHLIGHTED SUSTAINABLE PRACTICES:

Greywater reuse

Active and passive rainwater harvesting

THE SITE: Located on 3 acres in the Catalina Foothills of Tucson, Arizona, the site is draped over a narrow finger hillside, which has 12 feet (3.7 m) of vertical change and receives 12 inches (30.5 cm) of rain annually.

Project Description

Inspiration for this residential site comes from the history and living environment of the desert Southwest. The landscape highlights the richness of the many microclimates found in the Sonoran Desert and reuses greywater to support a diverse and xeric plant palette (see Figure 6.12). Initial discussions with the client, a native Australian accustomed to water shortages, established the project ethos to treat all water on-site as a precious resource.

D. A. HORCHNER (PHOTOGRAPHER)/DESIGN WORKSHOP, INC.

■ **FIGURE 6.12**
A sun-shade analysis of the site was used to determine the ideal placement of vegetation. The diverse and drought-tolerant plant palettes reflects the richness and beauty of the Sonoran Desert.

CATALINA FOOTHILLS *(CONTINUED)*

The landscape architect worked closely with the paving contractor to skillfully slope the pavement around trees and direct rainwater to the planting beds. A porous substrate under the paving passively stores rainwater where it can be easily accessed by tree roots (see Figure 6.13).

D. A. HORCHNER (PHOTOGRAPHER)/DESIGN WORKSHOP INC.

■ **FIGURE 6.13**
The stone patio is carefully graded to direct stormwater runoff to the tree basins. Porous substrate under the patio passively stores rainwater, where it is easily accessed by tree roots.

The architect and landscape architect coordinated construction plans to capture all available greywater sources and back-flush from the pool system. The opportunity to implement a greywater reuse system was timely, as Arizona had just become one of the first U.S. states to create a greywater reuse ordinance. This system was the first to be applied in a residential scenario in the Tucson area.

Rainwater harvested from the roof is stored and reused with the greywater. The combined water filters through a sediment and grease trap prior to going to the holding tank. The tank includes a final ultraviolet-light treatment and is connected to a conventional irrigation system that is pressurized by a small pump. The rainwater helps to dilute any residual salts. Nonphosphate and biologically friendly soaps and cleaning agents are used in the home to support the use of the greywater for irrigation.

PROJECT TEAM

- **DESIGN WORKSHOP, INC., LANDSCAPE ARCHITECT**
  Faith Okuma
  Claudia Meyer-Horn
  Sergio Yamada
  Wilbert Trujillo
  www.designworkshop.com
- **SUBY BOWDEN + ASSOCIATES, ARCHITECTS**
  www. sb-associates.net
- **WILLMENG CONSTRUCTION, CONTRACTOR**
  www.willmeng.com

Reclaimed Water Reuse

Reclaimed water, also known as recycled water, is wastewater that is treated and purified at a water treatment facility to standards that permit its safe reuse. Properly implemented reclaimed-water projects can assist communities in meeting water demands and solving supply challenges without any known significant public health risks (U.S. EPA 2004). The water does not have an odor nor does it stain sidewalks or other materials.

Common uses for reclaimed water include:

- Irrigation of residential lots, golf courses, playgrounds, sports facilities, orchards, and other agricultural fields. Unlike greywater, reclaimed water can be used in aboveground sprinkler systems.
- Fire protection
- Industrial purposes such as cooling and process water
- Toilet flushing
- Ornamental landscape water features not intended for human contact, such as decorative water fountains or reflecting pools
- Augmentation of surface water for downstream potable water reuse. Reclaimed water is often discharged to a water body and mixed with surface water, where it is purposefully reused downstream or in another location as a raw water supply for another water treatment plant.

Landscape irrigation is the second largest use of reclaimed water in the United States (Asano et al. 2006). The irrigation of public and private landscapes with reclaimed water is becoming more common in both arid and temperate climates (see Figure 6.14). For more than twenty years, the city of Tuscon, Arizona, has provided reclaimed water to approximately nine hundred sites, including golf courses, homes, parks, and schools. In 2009, the use of recycled water saved the city of Tuscon 5.5 billion gallons (20,819 megaliters) of drinking water, enough for 59,000 families for a year (Tucsonaz.gov 2010).

■ FIGURE 6.14 Community gardens irrigated with reclaimed water at the Mueller mixed-use development in Austin, Texas.

HEATHER VENHAUS

Reclaimed water systems are typically constructed, operated, maintained, and managed using methods similar to those of potable water systems (U.S. EPA 2004) and commonly consist of a connection to the reclaimed water-distribution main, a pipeline to the reuse area, a shutoff valve, and a flowmeter (Asano et al. 2006). Customers develop an agreement with the local water authority to provide access to the reclaimed water supply. Purple is the universal color of reclaimed water and is used for all the equipment and piping (see Figure 6.15).

■ **FIGURE 6.15** Purple irrigation equipment indicates reclaimed water use.

HEATHER VENHAUS

■ DESIGN CONSIDERATIONS

Reclaimed water is one of the few water supplies that will continue to grow as populations and community water demands increase. The water source is reliable year-round and in times of drought. In many areas, the use of reclaimed water not only benefits the site but also helps the community by reducing potable water demands and reducing or eliminating wastewater discharge to sensitive aquatic environments. However, reclaimed water may not be the most sustainable option due to the large amount of energy required to pump and treat it. By contrast, other alternative on-site water sources, such as harvested rainwater or greywater, require little to no energy use for most applications. Because of its large energy demand, the use of recycled water should be considered only after it is determined that alternative on-site water resources cannot meet site needs.

Regulations and guidelines for utilizing reclaimed water are not consistent between regions nor is the resource available to the public in all locations. The United States and Japan are the largest users of reclaimed water; more than half of U.S. states have water reclamation facilities (U.S. EPA 2004). Project teams should check with local water authorities to determine the availability and potential uses of this water resource. Reclaimed water rates are often lower than those of potable water and may provide substantial savings over the life of the project.

Public education to help site users become comfortable with the use of reclaimed water may be required. Reclaimed water has been researched extensively, and there are numerous examples of such water being used safely and successfully to assist the education process. Local codes may require signage to notify site visitors of the use of recycled water.

As with all water sources, the quality and potential impacts that reclaimed water may have on soil and vegetation must be understood. Water reclamation facilities and local governments should be able to provide water-quality information. The most important factor in determining the suitability of recycled water for irrigation is salinity, which can harm soil structure and vegetation (U.S. EPA 2004). Nutrients found in reclaimed water may be beneficial to the vegetation but, depending on site conditions, may not be necessary and may threaten nearby aquatic ecosystems. Low-profile sprinklers, microsprinklers, and drip irrigation systems can reduce foliar damage to trees and shrubs by limiting the contact of the reclaimed water with the leaves (Asano et al. 2006).

Advise all landscape and irrigation personnel of any recycled water restrictions and requirements. Important items to communicate in the site-monitoring and maintenance plan include:

- Regulations and use requirements, including the agreement with the water authority.
- Descriptions and details of the reclaimed water system, including guidance on how to properly operate the system.
- Monitoring and maintenance schedules: the reclaimed water system should be checked regularly to ensure it is functioning properly and for such issues as leaks, breaks, or clogs.
- Human health and safety measures: Recommend irrigation times when evapotranspiration rates are low and the potential for human contact is minimal. Describe the site conditions when reclaimed water should not be applied.
- Guidance on how to identify and mitigate damage to vegetation or soil that may be caused by the use of reclaimed water.

RESOURCES

Asano, T., F. Burton, and H. Leverenz. 2006. *Water reuse: Issues, technologies, and applications.* New York: McGraw-Hill.

Air-Conditioner Condensate Catchment and Reuse

A natural by-product of air-conditioning systems is condensate water. Similar to water collecting on the outside of a cold glass, condensation occurs when ambient water vapor—or humidity—comes in contact with the cooling components of air-conditioning equipment. Condensate is collected and removed from the air-conditioning unit to prevent damage and is commonly treated as a waste product that is disposed of in the sewer system. In order to collect and reuse condensate, the water is simply piped and gravity fed or pumped, either directly to the landscape or to a storage cistern where it can be used for irrigation or other landscape purposes (see Figure 6.16).

© MICHAEL KNOX

■ **FIGURE 6.16** This stylized sinkhole at the City Hall Plaza in Austin, Texas, demonstrates how rainwater enters the local aquifer. Condensation from the City Hall air-handler system is used to supply the water feature.

Air-conditioner condensate is essentially distilled water; it does not contain chlorine, minerals, or other additives, making it an excellent water source for irrigation. The water is low in suspended solids, turbidity, and salinity and has a pH that is neutral to slightly acidic (Kinkade-Levario 2007). The generation of condensate can be timely, because it typically coincides with the warm summer months, when irrigation demands are the greatest. The amount of condensate generated is largely dependent upon the local climate, building use, and the air-conditioning system. An estimated 3 to 10 gallons of condensate is generated per day per 1,000 square feet of air-conditioned space (11.35 to 37.84 L per day per 92.9 meters^2) (Alliance for Water Efficiency 2010). Buildings with high indoor/outdoor foot traffic and/or high occupancies tend to produce higher volumes of condensate because of the greater demands on their air-conditioning systems.

■ DESIGN CONSIDERATIONS

An integrated design approach that utilizes the professional expertise of building and landscape professionals is required to optimize the benefits of an on-site condensate-recovery system. Planning for condensate collection, storage, and reuse should begin early in the design process and is easier with new construction. Retrofitting existing buildings may be a bit more challenging; however, creative and cost-effective solutions can always be found (Bryant and Ahmed 2008).

Rainwater harvesting and condensate-recovery systems use similar tanks for storage and can be combined to increase efficiency and to reduce costs. Furthermore, supplementing collected rainwater with condensate provides a reliable and relatively steady source of water when rainwater supplies may be low or are unavailable. The combination of the two systems is sometimes referred to as "rainwater plus."

Air-conditioning condensate recovery is best suited for sites with large air-conditioned buildings located in hot and humid climates. Most single-family residential cooling systems are unlikely to provide significant quantities of condensate water; however, multifamily, commercial, industrial, and institutional buildings with large cooling demands can produce considerable volumes (see Figure 6.17).

Condensate may contain algae and other contaminants, such as heavy metals picked up from the air-conditioning equipment, that could make it unsafe for drinking. In addition to landscape purposes, the condensate can serve a variety of uses, such as decorative fountains and water features; however, the lack of minerals in the water makes it corrosive to most metals—particularly steel and iron (Alliance for Water Efficiency 2010).

BILL TIMMERMAN / PHOTOGRAPHER

■ **FIGURE 6.17**
The Underwood Family Sonoran Landscape Laboratory at the University of Arizona relies on recycled water from the site, which is comprised of approximately 40 percent condensate, 33 percent rainwater runoff, 18 percent well water blowoff, and 9 percent greywater.

The monitoring and maintenance plan should include as-built drawings of the condensate-recovery system as well as inspection schedules to ensure its proper function. Mechanisms for tracking the volume of condensate applied to the landscape should be incorporated into the system's design. Monitoring water use will help avoid waste and identify any leaks or system malfunctions. Maintenance staff are more likely to overirrigate with potable water, thereby defeating any water conservation measures, when they are not aware of the condensate volume being applied to the landscape.

■ RESOURCES

BuildingGreen.com. "Air Conditioner Condensate Calculator": http://www.buildinggreen.com/calc/calc_condensate.cfm

Kinkade-Levario, H. 2007. Design for water: *Rainwater harvesting, stormwater catchment, and alternative water reuse.* Gabriola Island, Canada: New Society.

Drought-Resistant Soils and Vegetation

In most regions of the world, droughts are a common and natural occurrence that are to be expected. However, thoughtful soil management and vegetation selection can reduce water requirements and buffer a site from drought conditions. Strategies for creating drought-resistant sites include the following:

- Avoid bare soils.
- Restore and maintain appropriate soil organic matter content.
- Recharge groundwater supplies.
- Select vegetation whose water demands can be fully met by the precipitation and the nonpotable water resources of the site.

AVOID BARE SOILS

Sunlight and wind can quickly evaporate moisture from bare soils. Mulch can be used to provide shade and prevent crusting of the soil surface, buffer temperature extremes, protect the soil from compaction, and reduce the evaporation of water.

Mulch falls into two basic categories: organic and inorganic. Inorganic mulches, including materials such as lava rock, mineral rock, gravel, and recycled glass, do not readily break down or provide organic material to the soil (see Figure 6.18). Materials derived from plants, such as compost, pine needles, leaves, straw, pecan hulls, and wood products, can be used as organic mulch (see Figure 6.19). Organic mulches decompose over time and are a good source of soil organic matter.

HEATHER VENHAUS

■ FIGURE 6.18
Recycled glass mulch adds interesting color and texture to the landscape and diverts waste from the landfill.

■ DESIGN CONSIDERATIONS

To protect overall soil health and conserve soil moisture, bare soils should be avoided in all project phases. Mulch is relatively inexpensive and versatile material. In addition to protecting soils, it can also be used to create site features, such as pathways and padded play areas.

Mulches derived on-site or from local resources can greatly reduce environmental impacts due to lower transportation requirements. Because mulch is the by-product of materials that commonly become organic waste—such as Christmas trees, land clearing, or tree trimmings—using mulch can divert "green waste" materials from the landfill.

Inorganic mulches are typically long-lived and need to be replaced less frequently than organic mulches. Like all materials, mulch can impact the site user's comfort as well as the site's microclimate. For example, dark stone mulches can absorb heat and increase surrounding temperatures, whereas light-colored mulches reflect sunlight, which may create an uncomfortable glare.

Applying mulch to landscapes is a widely accepted practice; if done incorrectly, however, it can damage plants and reduce the amount of water entering the soil. One common mistake is to apply mulch too deeply and too close to the base of vegetation. This can prevent water from entering the soil and cause plants to rot. Proper mulching practices should be outlined in the site maintenance and operation plan.

HEATHER VENHAUS

■ **FIGURE 6.19** Pecan hull mulch is a good source of organic matter and facilitates the reuse of an agriculture waste product.

MAINTAIN OR RESTORE APPROPRIATE ORGANIC MATTER CONTENT

Soil is a natural water reservoir that can sustain vegetation in times of drought. The amount of water in the soil that is available for plant use is largely determined by soil texture (i.e., the proportion of sand, silt, and clay), organic matter content, and the soil structure (i.e., the arrangement of the soil particles into aggregates). Altering soil texture can be difficult and resource-intensive due to the large amounts of sand, silt, or clay that must be harvested, transported, and integrated into the soil. A more sustainable and cost-effective approach for obtaining optimal soil conditions for drought resistance is to maintain—or if needed, increase—the amount of organic matter in the soil by using locally available compost.

Soil organic matter improves soil structure and helps to maintain pore spaces within the soil that hold air and water. Highly decomposed organic matter functions like a sponge and can absorb six times its weight in water. The amount of organic matter in a soil is determined by the climate, the vegetation, and the soil's location in the landscape, as well as the maintenance and management of the site. Soil organisms decompose organic matter, transforming it into nutrients and other substances beneficial to vegetation. Because organic matter continually decomposes and is not constant, it must be regularly replenished. Urban soils are commonly low in organic matter (Urban 2008) due to loss of vegetative cover and construction and maintenance practices that compact the soil, speed decomposition, and remove discarded plant materials from the site. Chapter 5, "Sustainable Solutions: Urban Flooding and Water Pollution," contains a broader discussion of organic matter and the strategies for maintaining or restoring soil organic matter content.

RECHARGE GROUNDWATER SUPPLIES

Water stored beneath the earth's surface in the pores and fractures of soil and rock is known as groundwater. As part of the hydrologic process, water infiltrates into the soil, recharging groundwater supplies, where it can be accessed by plant roots; flow underground into lakes, rivers, and other water bodies; or return to the surface via springs or man-made wells. Groundwater can remain stored within the soil and rock for extensive periods of time, serving as a natural buffer against drought conditions.

Site development and mismanagement can interfere with, and in some cases prevent, the recharge of groundwater supplies, resulting in a disruption of the hydrologic process and increasing the severity of droughts. Site strategies for improving groundwater recharge include:

- Minimizing impervious cover
- Restoring compacted and/or degraded soils
- Slowing stormwater runoff and increasing infiltration
- Mulching and shading soils to reduce evaporation

■ CASE STUDY

ROOGULLI GARDEN

PROJECT TYPE: Single-family residential

LOCATION: Bywong, New South Wales, Australia

SIZE: 2,153-square-foot (200 m^2) garden on a 25-acre (10-hectare) property

COMPLETION DATE: Begun in 2004 and still under construction in 2010

HIGHLIGHTED SUSTAINABLE STRATEGIES:

- Food production
- Reuse of on-site materials
- Use of local reclaimed materials
- House and landscape depend solely on harvested rainwater

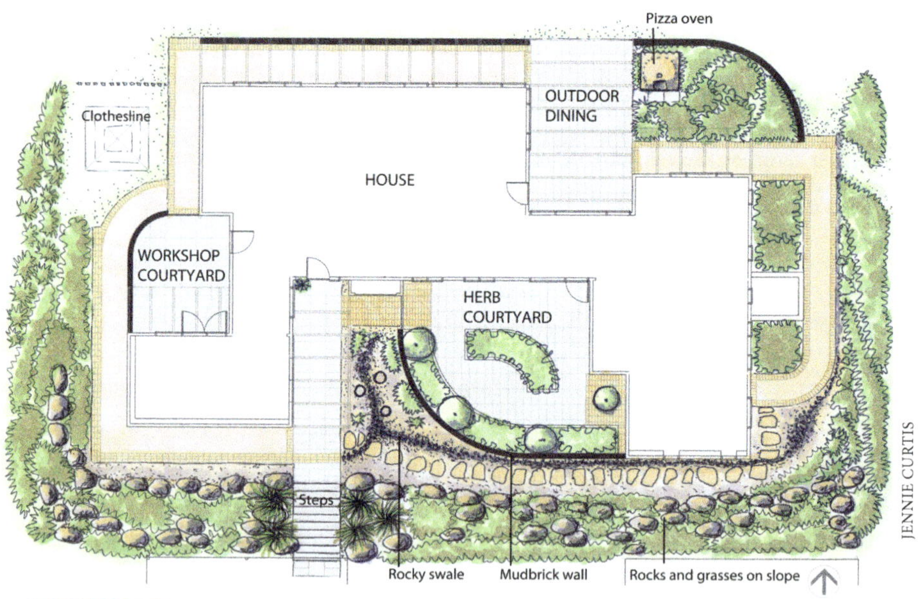

■ **FIGURE 6.20**
Roogulli garden master plan.

THE SITE: Situated close to Canberra, the site is on the southern tablelands of eastern Australia, an area known for its hot, dry summers and cold winters. The average annual rainfall for the area is 24 inches (600 mm); however, the region is currently recovering from a ten-year drought, during which average rainfall was reduced to around 16 inches (400 mm). Prior to construction, the site was degraded grazing land. The original grassy woodland ecosystem has been substantially modified, with most of the trees removed and exotic grasses introduced for agricultural purposes. Broad-scale application of fertilizers had raised the soil pH, and erosion was an issue. Soil salinity from the rising groundwater levels had also produced bare patches of soil.

ROOGULLI GARDEN *(CONTINUED)*

Design Overview

This residential garden, built by the homeowner, wraps around an energy-conscious home that relies solely on water from a 23,775-gallon (90,000 L) rainwater-harvesting cistern. Inspiration for the landscape was drawn from the Australian landscape, permaculture, and other xeriscape gardens. The slow process of building the garden has allowed plenty of time for the collection of secondhand materials and experimentation with new ideas (see Figure 6.20).

The design uses salvaged and recycled materials, creates opportunities to test new cultivars of Australian grassland plants, and supports a vibrant garden without artificial irrigation. It also provides plenty of food for the family and beautiful views from every window. Apart from the vegetable garden, the developed landscape is constrained to the area surrounding the house that was disturbed during construction. The garden has flourished with no artificial irrigation through many years of drought.

The house recedes into the landscape, and a curved mud-brick wall was added to provide a sense of enclosure. The north-facing wall in the courtyard acts as a thermal mass that creates a warm microclimate and helps to support citrus plants during the winter.

It became apparent that using salvaged materials requires more designer input during construction, and the design has adapted over the six-year construction period in response to the many opportunities discovered. One example is the shale gravel left over from sieving dirt for mud bricks, which was used in the swale near the front entry to solve drainage issues (see Figure 6.21).

PHOTO: KATINA CURTIS; LANDSCAPE ARCHITECTURE; JENNIE CURTIS

■ **FIGURE 6.21**
Mud-brick wall, dry creek drainage swale, and recycled pavers.

Stone, fired bricks, timber sleepers, concrete pavers, and clay pavers were salvaged from other sites. Gaps between the rock retaining wall on the south side of the house were filled with soil from the site and planted with Australian grasses, lilies, and rushes (see Figure 6.22). The pizza oven is also constructed from soil found on-site and finished with an earth paint. Mulch is either shale gravel from the site or shredded green waste from the local tip. Many of the decorative elements are made from salvaged items, including fencing wire and an old cooking pot.

continues

ROOGULLI GARDEN *(CONTINUED)*

PHOTO: KATINA CURTIS;LANDSCAPE ARCHITECTURE; JENNIE CURTIS

In the vegetable garden, heirloom fruits and vegetables are grown with compost made on-site using manure from alpacas raised on the site. As well as providing eggs, the hens are used to control insect pests and clean up old plants in the vegetable garden (see Figure 6.23). This is a low-technology garden with a great deal of embodied human energy.

■ **FIGURE 6.22**
Rock retaining wall, Australian grass tree (*Xanthorrhoea sp.)*, and other native grassland plants.

■ **FIGURE 6.23**
Chickens control insect pests in the vegetable garden.

PROJECT TEAM

JENNIE CURTIS,
FRESH LANDSCAPE DESIGN,
LANDSCAPE ARCHITECT
www.freshlandscape.com.au

Peter Adamson, Architect

www.peteradamsonarchitects.com.au

Chris and Jennie Curtis (owners),
Out and About Landscapes,
Construction

PHOTO: KATINA CURTIS; LANDSCAPE ARCHITECTURE; JENNIE CURTIS

SELECT VEGETATION WHOSE WATER DEMANDS CAN BE FULLY MET BY THE PRECIPITATION AND THE NONPOTABLE WATER RESOURCES OF THE SITE

Vegetation is a central element of site design that not only provides aesthetic value but also a variety of ecosystem services such as soil stabilization, air and water cleansing, and climate regulation. However, when plants are not well suited to a site, and require regular potable water irrigation, they can create a host of problems that can outweigh the benefits. Environmental issues associated with potable water stem from not only freshwater shortages but also the substantial energy requirements and greenhouse gas emissions associated with the treatment and transport of the water. To avoid this unsustainable scenario, plants selected should match landscape conditions and be fully sustained by precipitation or alternative on-site water resources.

All too often, projects near completion before the site's water requirements are fully understood. Delaying water-conservation efforts until late in the design process leaves little time for adjustment and often results in missed opportunity to avoid water waste.

The design process described below is adapted from xeriscaping principles and outlines steps a project team can take to create a planting plan that eliminates potable water demands and wisely utilizes the site's alternative water resources.

1. Investigate site characteristics, such as topography and soils, to gain an understanding of the growing conditions and movement of water across the site.
2. Conduct a water-balance analysis to determine the quantity of water available each month from precipitation and alternative water resources, such as harvested rainwater, greywater, and air-conditioner condensate. The use of potable water should be reserved for extreme drought conditions or during the establishment phase, when additional irrigation may be required.
3. Establish the size of the irrigated area based on the site conditions and available nonpotable water budget. Divide the irrigated area into zones and establish the amount of water available to each zone on a monthly basis. Depending on the quantity of water available, some areas of the landscape may need to be designed without supplemental irrigation. Site conditions and the program plan should determine the size and location of the vegetated areas.
4. Select plants for each zone according to their resource requirements. Consider the mature size of the vegetation and avoid the temptation to overplant for immediate gratification. Overplanting results in increased water demand and pruning practices.
5. Design the irrigation system to individually address the unique water requirements of each zone. Depending on the quantity of the water supply, not all areas may receive irrigation.
6. Use the site maintenance plan to educate caretakers about irrigation zones, their water requirements, and appropriate irrigation methods.

Working through this process and gaining a clear understanding of the site's water availability often encourages the design team to explore sustainable strategies that increase on-site water supply and improve the drought resistance of the landscape. Conducting the first three steps early in the design process will allow time for exploration and adjustment. The steps may need to be repeated multiple times as different design options are evaluated.

■ DESIGN CONSIDERATIONS

The success and health of vegetation largely depends on the condition of the soil in which it grows. Prior to planting, degraded soils need to be restored. In order to encourage infiltration and the storage of water, special attention should be given to the soil's bulk density and organic matter content.

A significant portion of the water used in landscapes is applied to turfgrass, which is commonly overwatered (Rain Bird 2004). The design team should consider when and how turfgrass or other water-intensive vegetation will be used, both so it can be sized according to its practical function and so suitable drought-tolerant alternatives can be identified.

Materials and plants purchased for the site should embody the sustainable goals of the project. The design team should research potential manufacturers and plant nurseries to gain a better understanding of how their materials are extracted, built, grown, and transported. Plants should be purchased from plant suppliers who know the source of their materials and employ environmentally friendly practices such as integrated pest management, composting, the use of renewable energy sources, and water-conservation practices.

Plants usually require supplemental water during the establishment phase—typically one to three years, depending on the vegetation type and location. Additional water used during establishment may temporarily increase the site's water usage; however, over the long term, it will reduce plant mortality and replacement costs. Irrigation practices during the establishment phase should be tailored to the specific vegetation type and focus on irrigating in a fashion that conserves water and encourages deep-rooted vegetation. Because establishing vegetation typically requires more water, preserving existing, established, healthy plants saves water by reducing the need for supplemental irrigation of new plants (U.S. EPA 2002).

Maintenance practices can reduce the drought tolerance of a landscape. For example, fertilization encourages rapid growth, which requires additional water, and vehicles or heavy equipment can compact the soil, limiting infiltration and water-holding capacity. The site monitoring and maintenance plan should specify strategies and schedules that save water while also nurturing the beauty of the site. A contingency plan describing the conditions under which it is appropriate to use potable water on the landscape, and the necessary quantity, should also be included in the maintenance plan.

■ RESOURCES

Native Plant Information Network at the Lady Bird Johnson Wildflower Center: www.wildflower.org

U.S. Department of Agriculture (USDA). Natural Resources Conservation Service. Plants Database: http://plants.usda.gov/java/

DROUGHT-TOLERANT VEGETATION

Site planners who are interested in conserving water are not limited to rock and cacti (U.S. EPA 2002). A diverse selection of colorful and lush drought-tolerant plants are available through the nursery trade. It is a common misconception that native plants are inherently drought tolerant. As with all vegetation, the growing requirements of native plants vary and are dependent upon the ecosystem in which the plant originated.

Characteristics that typically indicate drought tolerance include the following:

- Small or divided waxy or hairy leaves that reduce water loss due to transpiration
- Wide-spreading surface root systems that quickly absorb rainfall
- Deep-rooted vegetation that uptakes water from deep within the soil profile
- The ability to drop leaves during times of drought and quickly regrow new leaves when environmental conditions improve

- Plants that are native to landscapes that are arid, experience frequent drought, or have soils with low water-holding capacity
- Grey or white foliage that helps reduce water loss
- Succulent leaves that can store water
- Aromatic foliage

Not all drought-tolerant vegetation exhibit these characteristics. Local nurseries and plant experts should be consulted for a thorough list of drought-tolerant vegetation appropriate for specific site conditions and project requirements (see Figure 6.24).

■ **FIGURE 6.24** Low-water-use garden in the foothills of the Colorado Rocky Mountains (a). The design, developed and installed by Lauren Springer Ogden, is plant-driven and relies on vegetation to vividly express form, texture, and year-round appeal (b).

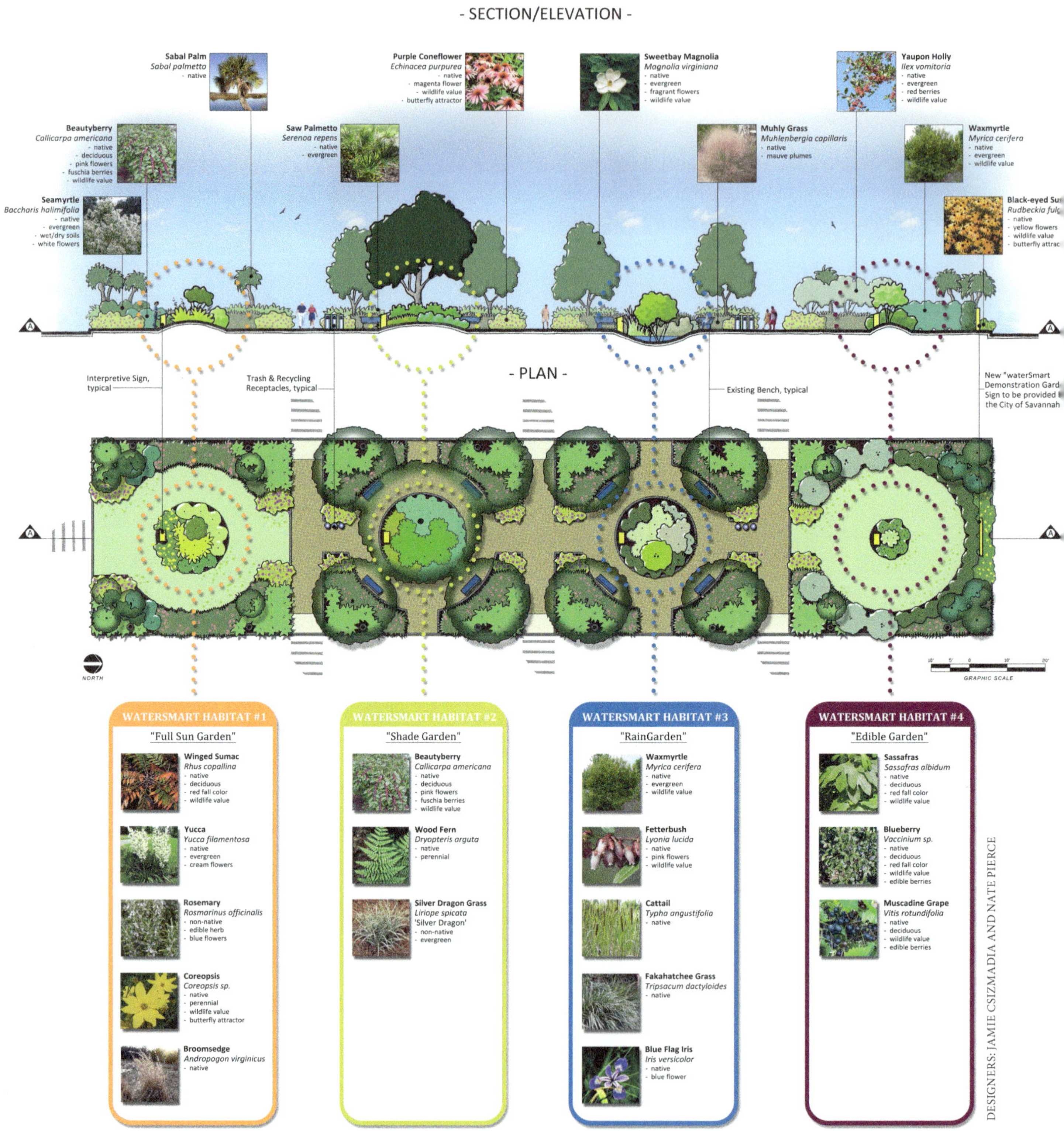

DESIGNERS: JAMIE CSIZMADIA AND NATE PIERCE

■ **FIGURE 6.25**

The Bryan Square WaterSmart Demonstration Garden assists Savannah, Georgia, residents in creating landscapes that are beautiful and can be maintained with little or no supplemental watering.

Avoid Wasteful Irrigation Practices

Conventional irrigation practices commonly waste a significant amount of water due to evaporation, runoff, and overwatering (U.S. EPA 2002). The design of the irrigation system, as with that of all other site components, should be an integrated process that responds to the unique character of the site and the requirements of the program plan. Throughout the design process, the integrated design team should explore options to use alternative water supplies and eliminate potable water consumption.

Irrigation systems are not limited to potable water supplies and can be connected to alternative water sources, such as rainwater and air-conditioner condensate cisterns. In response to global water shortages, irrigation companies are becoming leaders in water conservation and are continually developing new technologies to reduce water use. Water-efficient irrigation practices have the potential to significantly decrease the water use of landscapes while sustaining the health and beauty of vegetation (Rain Bird 2004).

Water-efficient irrigation practices include the following:

- Avoid irrigating on a regular schedule that does not reflect current site conditions or plant requirements.
- Divide the landscape into separate irrigation zones according to the water requirements of the vegetation.
- Water vegetation slowly and deeply. Apply water in two or more short cycles to encourage infiltration and reduce runoff.
- Avoid watering in sunny or windy conditions that increase evaporation. Early morning watering is typically the most efficient.
- Use low-volume irrigation devices such as bubblers and drip emitters, which deliver water slowly at or near the plant base.
- Specify smart irrigation control systems that account for conditions such as weather, evapotranspiration, soil moisture, and plant type to determine when irrigation needs to occur, rather than relying on a preset schedule.
- Regularly check irrigation equipment for leaks and breaks. Adjust irrigation equipment and schedules to address changes in the landscape and seasons. Monitoring the amount of water used by the irrigation system will help flag any leaks or breaks within the system and will encourage conservation efforts.

■ DESIGN CONSIDERATIONS

As with all technologies, water-conserving irrigation systems and devices must be installed and operated correctly in order to reduce water use effectively. Systems need to be checked to ensure proper installation and efficiency standards prior to finalizing the construction process.

The site monitoring and maintenance plan should describe the plant and soil conditions that indicate when additional irrigation is necessary. This information replaces regular watering schedules or amounts that may not take into consideration the current weather, season, or plant conditions at the site. Site caretakers should also receive guidance on how to use irrigation systems efficiently and check for leaks or other failures.

■ RESOURCES

Cross, R., and R. Spencer. 2008. *Sustainable gardens.* Collingwood, Victoria, Australia: CSIRO.

U.S. Department of Interior. Bureau of Reclamation. Lower Colorado Region. Southern California Area office. 2007. *Reclamation: Managing water in the West. Weather and soil moisture–based landscape irrigation scheduling devices: Technical review report, 2nd ed.*

REFERENCES

Alliance for Water Efficiency. 2010. "Condensate water introduction." Retrieved November 12, 2010, from http://www.allianceforwaterefficiency.org/Condensate_Water_Introduction.aspx.

Arizona Department of Water Resources [ADWR]. 2011. "Residential home page." Retrieved March 1, 2011, from http://www.adwr.state.az.us/azdwr/StatewidePlanning/Conservation2/Residential/Residential_Home2.htm.

Asano, T., F. Burton, and H. Leverenz. 2006. *Water reuse: issues, technologies, and applications.* New York: McGraw-Hill.

Bryant, J., and T. Ahmed. 2008. Condensate water collection for an institutional building in Doha, Qatar: An opportunity for water sustainability. Paper presented at the 16th Symposium on Improving Building Systems in Hot and Humid Climates, Plano, TX.

Cross, R., and R. Spencer. 2008. *Sustainable gardens.* Collingwood, Victoria, Australia: CSIRO.

European Environmental Agency [EEA]. 2010. "The European environment—state and outlook 2010: Use of freshwater resources." Retrieved April 1, 2011, from http://www.eea.europa.eu/data-and-maps/indicators/use-of-freshwater-resources/use-of-freshwater-resources-assessment-2.

Fuller, T. 2010. "Countries blame China, not nature, for water shortage." *New York Times*, April 1.

Drought has also created a major public relations problem for the Chinese government in neighboring countries, where, in recent years, China has tried to project an image of benevolence and brotherhood.

Government of South Australia. South Australia Water. 2004. "Mawson Lakes recycled water system." Retrieved November 6, 2010, from http://www.sawater.com.au/SAWater/WhatsNew/MajorProjects/mawson_lakes.htm.

Hoekstra, A. Y., and A. K. Chapagain. 2006. Water footprints of nations: Water use by people as a function of their consumption pattern. *Water Resources Management.*

Kinkade-Levario, H. 2007. *Design for water.* Gabriola Island, Canada: New Society.

Lancaster, B. 2004. *Rainwater harvesting for drylands and beyond, vol 1: Guiding principles to welcome rain into your life and landscape.* Tucson: AZ: Rainsource.

Ludwig, A. 2009. *Create an oasis with greywater: Choosing, building, and using greywater systems.* Santa Barbara, CA: Oasis Design.

Matalon, L. 2010. Colorado River water rights. *The World,* April 9. http://www.theworld.org/2010/04/colorado-river-water-rights/.

Rain Bird. 2004. *Irrigation for a growing world.* Azusa, CA: Rain Bird Corporation.

Roesner, L., Y. Qian, M. Criswell, M. Stromberger, and S. Klein. 2006. *Long-term effects of landscape irrigation using household graywater: Literature review and synthesis.* Alexandria, VA: Water Environment Research Foundation; Washington, DC: Soap and Detergent Association.

Texas Water Development Board. 2005. *The Texas manual on rainwater harvesting.* Austin.

Tucsonaz.gov. "Reclaimed water." Retrieved October 11, 2011, from http://cms3.tucsonaz.gov/water/reclaimed.

Urban, J. 2008. *Up by roots: Healthy soils and trees in the built environment.* Champaign, IL: International Society of Arboriculture.

U.S. Department of Interior. Bureau of Reclamation. Lower Colorado Region. Southern California Area office. 2007. *Reclamation: Managing water in the West. Weather and soil moisture–based landscape irrigation scheduling devices: Technical review report, 2nd ed.*

U.S. Environmental Protection Agency [U.S. EPA]. 2004. *Guidelines for water reuse.* Washington, DC: U.S. EPA, Office of Water Management.

U.S. General Accounting Office [U.S. GAO]. 2003. *Freshwater supply: States' views of how federal agencies could help them meet the challenges of expected shortages*: 110.

Watkins, K. 2006. *Human development report 2006: Beyond scarcity: Power, poverty and the global water crisis.* New York: United Nations Development Programme.

World Water Assessment Programme [WWAP]. 2009. *The United Nations world water development report 3: Water in a changing world.* Paris, London: UNESCO / Earthscan.

ROB HAINER/BIGSTOCK.COM

■ **FIGURE 7.1**
Introduced to the United States in 1876 as an ornamental and propagated widely for livestock forage and erosion control, Kudzu (*Pueraria montana*) now dominates over 7 million acres across the southern United States.

CHAPTER 7

Sustainable Solutions: Invasive Species

■ BY W. MATT MCCAW

A FOUNDATIONAL PRINCIPLE of an ecological education is the notion of a species' native status. The idea has to do with where a species evolved and was able to establish without the aid of humans. As an example, honey mesquite (*Prosopis glandulosa*) occurred in the semiarid plains of the southern United States and northern Mexico at the time of European settlement; thus, honey mesquite is native to the southern United States and northern Mexico. In the last few hundred years, it has expanded its native range in response to agricultural activity but is still, by and large, considered native in those areas. However, in the 1920s, mesquite was introduced to western Australia as a forage plant and as an ornamental tree that could withstand arid conditions. It subsequently escaped cultivation and is now considered one of the top twenty noxious weeds in Australia (Rangelands 2008) and one of the world's one hundred worst invasive species (Lowe et al. 2000).

An invasive species is defined as one that is nonnative to a particular ecosystem and whose introduction into that system causes or is likely to cause economic or environmental harm or harm to human health (Executive Order No. 13,112, 1999). The key to this definition is that an invasive species causes, or is likely to cause, harm. It is preemptive. We do not have to wait until a nonnative species causes harm before we treat it as invasive.

There are many indications that a nonnative species might become invasive in a new ecosystem:

- It has a history of invasion in other ecosystems.
- It has escaped cultivation in the new ecosystem.
- Its native ecosystem is similar to the new ecosystem.
- It grows aggressively, is weedy, and/or is highly abundant or dominant in its native ecosystem.

Unfortunately, there are no sure methods for predicting invasions. The only method for preventing invasions is the exclusive use of native or sterile nonnative species.

The scope of this chapter has been limited to plants because they are the most prominent invasive species in built landscapes. However, around the globe, known invasive species include fungi and all forms of both terrestrial and aquatic animals, including vertebrates, insects, and crustaceans.

Invasive Species: The Cause

Most nonnative plants now established in the United States were deliberately introduced for agricultural or ornamental purposes (Pimentel et al. 2005). Though over 20 percent of plant species in U.S. natural ecosystems are nonnative (Morin 1995), not every introduced species becomes invasive. A crude rule is that one in ten imported species becomes established in the wild, and of those, one in ten becomes a pest (Williamson and Fitter 1996). In Florida, more than nine hundred nonnative plant species, about 4 percent of those that have been introduced, have become established in natural ecosystems (Pimentel et al. 2005). The problem is that while only a minority of introduced plants becomes invasive, we cannot reliably identify these species ahead of time in order to prevent their introduction. In addition, the relative few that do become invasive cause significant damage to economies, the environment, and human health.

Most invasive species are habitat generalists—they are able to survive and reproduce in a variety of environments (Baker 1974). This contrasts with habitat specialists that perform very well, but only in a narrow range of environmental and ecological conditions. For example, high-diversity native grass-seed mixes have been shown to establish more quickly and provide earlier erosion control than low-diversity seed mixes comprised mostly of Bermuda grass (*Cynodon dactylon*) (Tinsley et al. 2007), which is invasive throughout the United States and is used widely for erosion control along roadsides. There is a tradeoff, then, between habitat breadth and competitive ability—that is, the "jack of all trades, master of none" model (Marvier et al. 2004). Generalists can persist, even thrive, in a wide variety of environmental conditions, but they are rarely the top competitors in healthy ecosystems. As a result, nonnative generalists do not typically invade and dominate intact communities in which there is robust competition among natives (Marvier et al. 2004; Daehler 2003).

It is true that invasive species can play a part in displacing native species, but invasive species are often maligned because of the notion that they outcompete natives on a fair playing field. This idea is usually invalid, at least during the early phases of an invasion. In order for a nonnative generalist to become a widespread dominant, competition among natives must usually be reduced (Daehler 2003; MacDougall and Turkington 2005). This alteration of the competitive dynamic among native species is most often wrought by human disturbances (Daehler 2003), such as overgrazing, soil erosion, irrigation and fertilization, certain types of mowing, and the disruption of natural processes such as fire and flooding.

Unfortunately, once nonnative species have invaded disturbed ecosystems, the cessation of human disturbance and the reintroduction of natural processes are rarely adequate for their control. Invasive species often establish positive feedback mechanisms (such as altered fire regimes, altered soil chemistry, and dominance of the seed bank) that further enforce their own dominance (Suding et al. 2004) or even facilitate the invasion of other nonnative species (Simberloff and Holle 1999). This is usually the point at which invasive species, on their own, have the potential to outcompete native species and reduce biodiversity over time. When this phase of the invasion is reached, the removal of invasive species and the restoration of the degraded systems often requires drastic and expensive measures (Suding et al. 2004).

Most exotic species invasions experience a time lag between their introduction and their explosion in population and invaded area. A plant may be used for years or decades in horticultural or agricultural settings before it exhibits indications of invasiveness. The size of the nonnative population affects the length of the time lag (Crooks and Soulé 2001); that is, the more an introduced plant is propagated, the greater the seed output, the higher the likelihood of invasion, and the more rapidly invasion will occur.

Only a minority of introduced species become invasive. However, of the thousands of invasive plant species now degrading ecosystems around the world, many were introduced as ornamentals. Here are a few examples:

United States

SALT CEDAR (*TAMARIX SPP.*)

Salt cedar (Figures 7.2 and 7.3) was introduced to the United States from Eurasia in the 1800s for its attractive flower, its low maintenance requirements, and as a means of preventing erosion along stream banks. By 1998 salt cedar had invaded essentially every drainage system in the arid and semiarid southwestern U.S. and western Mexico. It is a facultative phreatophyte, meaning that its deep roots are capable of drawing moisture from groundwater tables, when available. It also changes the soil's chemical profile by depositing salt both aboveground and belowground. The salt deposits inhibit other plants from growing and degrade native habitat.

STEVE DEWEY, UTAH STATE UNIVERSITY, BUGWOOD.ORG

■ **FIGURE 7.2**
Salt cedar flowers.

STEVE DEWEY, UTAH STATE UNIVERSITY, BUGWOOD.ORG

■ **FIGURE 7.3**
Salt cedar has invaded vast expanses of the southwestern United States and northern Mexico.

PURPLE LOOSESTRIFE (*LYTHRUM SALICARIA L.*)

Purple loosestrife (Figures 7.4 and 7.5), an herbaceous perennial native to Eurasia, was introduced to the northeastern United States and Canada in the 1800s for ornamental and medicinal uses. Loosestrife adapts readily to natural and disturbed wetlands, forming dense, homogeneous stands that displace native wetland plant species, including some U.S. federally endangered orchids, and degrades waterfowl habitat. It is estimated that loosestrife now occurs in forty-eight states across the United States and costs $45 million per year in control costs and forage losses.

LINDA HAUGEN, USDA FOREST SERVICE, BUGWOOD.ORG

■ **FIGURE 7.4** Purple loosestrife can degrade wetland diversity and aquatic habitat.

LINDA HAUGEN, USDA FOREST SERVICE, BUGWOOD.ORG

■ **FIGURE 7.5** Purple loosestrife in a residential flower bed. This species was introduced to North America via the horticulture trade and is still occasionally used in landscape plantings.

United Kingdom

COMMON RHODODENDRON (*RHODODENDRON PONTICUM*)

Common rhododendron, an evergreen shrub native to southern Europe and southwestern Asia, was introduced to the United Kingdom as an ornamental, as well as to provide cover for game birds. It forms dense thickets with deep shade that exclude native species. The cost of controlling common rhododendron in the UK has been estimated at $288 per acre (£526 per hectare) (Dehnen-Schmutz et al. 2003).

China

CANADA GOLDENROD (*SOLIDAGO CANADENSIS*)

Goldenrod (Figure 7.6) was initially introduced as an ornamental garden plant to Shanghai in 1935 and has since spread to over ten Chinese provinces. Goldenrod is an ornamental herbaceous perennial native to North America that creates dense monocultures that exclude native plant species. This results in the loss of plant and insect diversity and, ultimately, alteration of ecosystem function (Dong et al. 2006).

MRS. W. D. BRANSFORD, LADY BIRD JOHNSON WILDFLOWER CENTER

■ **FIGURE 7.6**
Canada goldenrod.

Invasive Species: How They Affect Our Lives

The global economic cost of invasive species has been estimated at US$1.4 trillion (Steiner 2010). Invasive species handicap the Canadian economy to the tune of between C$13.3 and C$34.5 billion (US$13.3 and US$34.5) each year (Colautti et al. 2006). The economic cost to the U.S. economy is estimated at $120 billion per year (Pimentel et al. 2005). In Germany, the control of black cherry (*Prunus serotina*) alone costs approximately €25 million per annum, and invasive ragweed and giant hogweed cause public health expenditures of €33.2 million per annum (Reinhardt et al. 2003).

Invasive species also degrade ecosystem services. In South Africa, exotic woody plants reduce available surface water by about 7 percent (Le Maitre et al. 2000). A program to bring the invasion under control would cost US$1.84 billion over twenty years (Le Maitre et al. 2002). Invasive plants degrade rangelands by reducing forage productivity, reducing yields, and poisoning livestock and cause an annual economic loss of over $2 billion in the United States (DiTomaso 2000).

Invasive species can contribute to extirpation or extinction. Forty-two percent of threatened or endangered species in the United States are imperiled at least partly because of invasive species (Pimentel et al. 2005). Worldwide, invasive species have been implicated in over half of all modern extinctions for which there is adequate data. It is believed that invasive species are the leading cause of extinction of birds and the second leading cause of extinction of fish and mammals (Clavero and Garcia-Berthous 2005).

Control Invasive Species and Prevent New Invasions

Human activity is responsible for the spread of invasive species, and only human activity will bring invasions under control. Design teams and site managers should employ the following strategies to control invasive species and prevent new invasions:

- Use only site-appropriate plant species.
- Create and maintain invasion-resistant plant communities.
  - Encourage high-diversity plant communities.
  - Minimize resource inputs.
  - Minimize bare ground.
  - Limit habitat fragmentation.
  - Maintain healthy disturbance.
- Use integrated pest management.
  - Identify pests.
  - Set action thresholds.
  - Prevent the establishment of new invasive species or the spread of those already established.
  - Control invasive species.
  - Monitor and follow up.

Use Only Site-Appropriate Plant Species

Site-appropriate species are noninvasive and, once established, are able to thrive without supplemental potable irrigation, fertilization, or significant fossil fuel–powered maintenance. The use of particular nonnative species is appropriate on some sites, but by and large the use of nonnatives is a gamble even if the species has no history of invasion in similar ecosystems. Essentially, the only guarantee that a nonnative will not become invasive is the biological inability of the varietal to produce viable seed. In some cases, however, sterile varietals are visually indistinguishable from their fertile conspecifics [e.g., the fertile varietal of heavenly bamboo (*Nandina domestica*) is visually identical to the sterile varietal until it bears fruit]. Often, the fertile varietal is purchased and planted by mistake. By the time the fertile varietal begins to produce seed, the project is finished and the design team has taken its botanical expertise elsewhere. At that point, the fertile invasive species usually remains long-term, cranking out viable seed and spreading to other sites.

Design teams should exercise an abundance of caution when using nonnative species. Remember, we cannot predict which nonnative species will become invasive, and nonnative species usually exhibit time lags between introduction and invasion. Furthermore, professionals who are relied upon by clients and customers for advice and guidance on plant selection have significant power in controlling the distribution and spread of invasive species. By using species that can potentially become invasive, landscape practitioners may be contributing to long-term damage to economies, the environment, and human health.

Create and Maintain Invasion-Resistant Plant Communities

Invasive species do not plow indiscriminately across the landscape. Their distribution, at least until a critical mass is reached, is primarily determined by human activity (MacDougall and Turkington 2005). Further, they establish in areas that provide favorable growing conditions and generally where competition among native species has been reduced (Daehler 2003). Thus, the keys to creating, restoring, and maintaining invasion-resistant plant communities are controlling the influx of propagules and encouraging and maintaining intense competition among desirable natives.

ENCOURAGE HIGH-DIVERSITY PLANT COMMUNITIES

There are many parallels between ecological and economic systems. One is that competition is intense when there are many players in the market. A robust market with many competitors is not likely to be dominated suddenly by a newcomer.

Not surprisingly, in a variety of settings, high native plant diversity has been shown to resist invasion by exotic species (Kennedy et al. 2002; Maron and Marler 2007). In addition, the diversity of functional groups (e.g, bunch grasses, sod grasses, annuals, perennials, cool-season and warm-season plants, etc.) may be as important to invasion resistance as the diversity of species (Maron and Marler 2007; Pokorny et al. 2005). High-diversity plant communities exhibit, as ecologists say, high niche complementarity (see Figure 7.7); these communities make use of much of the available resources (water, nutrients, and light) and leave very little for invaders. Oftentimes, high-diversity assemblages may dramatically reduce invasion or completely exclude nonnative species. The targeted reduction of certain invasive species can sometimes be achieved by the introduction of native species belonging to the same functional group as the invasive. In such cases, the success of the invasive species is reduced because the native habitat specialist competes more effectively for resources and also competes directly with the invasive species for those resources. Simmons (2005) significantly reduced the dominance of the invasive annual forb *Rapistrum rugosum* (Figure 7.8) by seeding a native annual forb, *Gaillardia pulchella*, with similar biomass production and phenology as the invasive (Figure 7.9).

CITY OF AUSTIN

■ **FIGURE 7.7** In diverse herbaceous communities, many plant species may occupy the same physical space at different times during the year. Here, native annual forbs—firewheel (*Gaillardia pulchella*), lanceleaf coreopsis (*Coreopsis lanceolata*), and black-eyed Susan (*Rudbeckia hirta*)—flower during a particularly wet spring in central Texas. In just a few weeks, these species will set seed and be replaced by a new procession of native forbs and grasses.

MARK SIMMONS; PHOTO COURTESY OF THE LADY BIRD JOHNSON WILDLIFE CENTER

■ **FIGURE 7.8** *Rapistrum rugosum* (yellow) dominates a roadside community of grasses and forbs. *R. rugosum* invades most aggressively when frequent mowing suppresses competition from native plants.

SIMMONS 2005

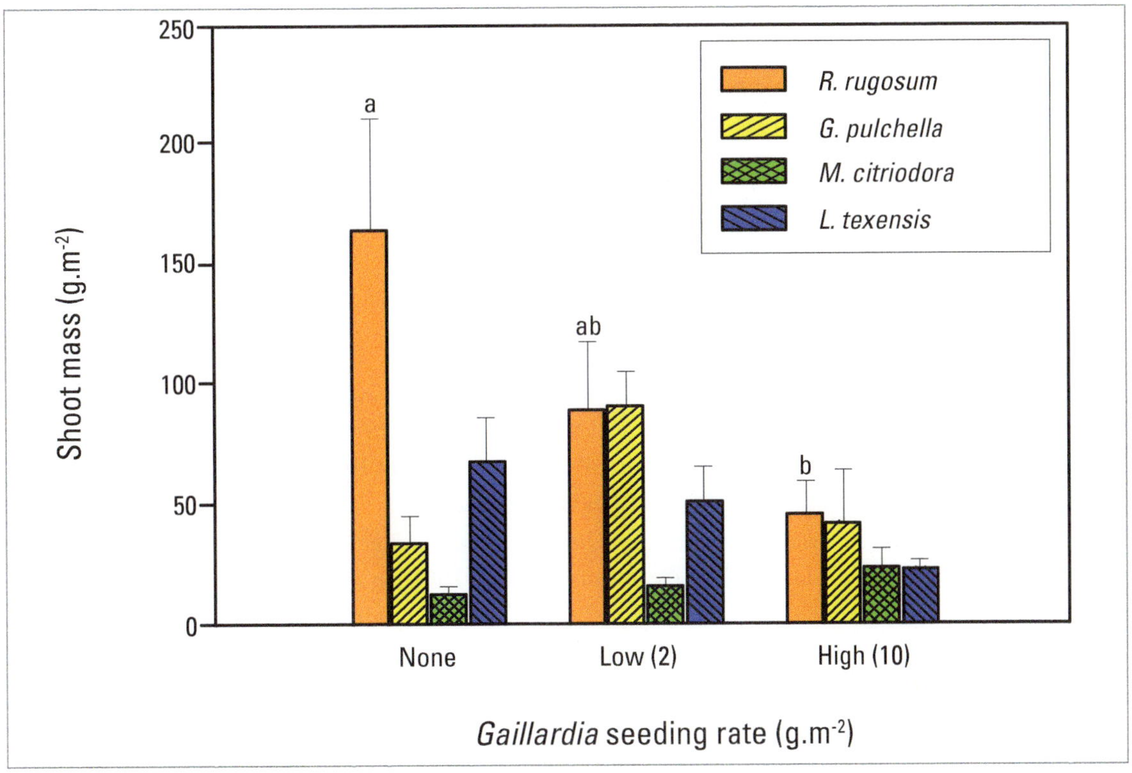

■ **FIGURE 7.9** Sowing native *G. pulchella* over established seedling colonies of *R. rugosum* resulted in significant reduction of *R. rugosum* productivity. The highest sowing rate of *G. pulchella* resulted in a 72 percent reduction in aboveground biomass of *R. rugosum* that translated to an estimated 83 percent decrease in seed set, without significant suppression of adjacent native species. The graph shows the effect of oversowing seeds at three different rates on the production of *R. rugosum*, *G. pulchella*, *Monarda citriodora*, and *Lupinus texensis*. Bars with different letters are significantly different at $\alpha = 0.05$ level.

MINIMIZE RESOURCE INPUTS

Another parallel between ecological and economic systems is that competition is intense when resources are scarce. The addition of resources via fertilization and irrigation reduces or suspends competition and increases invasibility of a plant community (Davis and Pelsor 2001). Irrigation and fertilization, most commonly used in grassland systems, have caused widespread declines in plant diversity, usually by promoting the dominance of perennial grasses to the detriment of forbs and annual grasses and by facilitating the invasion of exotic species (Hobbs and Huenneke 1992). Even short-term irrigation has been shown to encourage species invasions (Davis and Pelsor 2001). Wetlands are particularly susceptible to invasions because of reliable water availability, nutrient inputs, sediment deposition, and inflow of propagules. Twenty-four percent of the world's most invasive plants are wetland species, even though wetlands comprise only six percent of the earth's surface (Zedler and Kercher 2004).

Invasive species are generally not the best competitors. They typically have the potential for rapid growth and displacement of native species, but they require the availability of abundant resources to fuel that growth. When resources are scarce, invasive species may fail to establish; if they do establish, their performance may be poor and they may not achieve dominance in the community.

Minimizing irrigation and fertilization is the first step to managing resource availability. However, some sites, because of past land use, may contain high levels of soil nutrients such as nitrogen and phosphorus. Natural losses of artificially high soil nitrogen or phosphorus levels are usually too slow from a management perspective. To speed nutrient removal, site managers might employ cropping, which involves the cutting and removal of aboveground herbaceous biomass each season (Hobbs and Huenneke 1992). Soil carbon supplements such as sucrose, sawdust, and organic mulch may also be used to increase soil carbon-to-nitrogen ratios and increase microbial uptake of nitrogen and phosphorus. Carbon supplements in grasslands have been shown to decrease the dominance of invasive plants while improving the success of native species (Daehler 2003).

Other sites may have excess stormwater or greywater that needs to be managed on-site. This may present somewhat of a conflict between the need to manage on-site water resources and the need to improve invasion resistance. Invasibility can be reduced by applying stormwater or greywater to wetland or riparian vegetation, as in rain gardens or bioretention basins, that is suited to the augmented moisture regime. Minimizing soil disturbance and maintaining near–100 percent groundcover will discourage the establishment of newly arrived seeds of invasive plants. Diverse plant communities will be more productive vegetatively and will compete more efficiently for any light and nutrients.

MINIMIZE BARE GROUND

A potentially opportune time for a nonnative plant to invade a site is when a disturbance has just removed some of its competition. "Nature abhors a vacuum," goes the old adage. Leave the ground bare for too long, and nature will put something there. Occasionally, that something is desirable and we smile; other times that something is a pest, and we grimace.

It should come as no surprise that researchers have observed higher levels of invasion when bare ground is fertilized (Burke and Grime 1996; Thompson et al. 2001) or irrigated (Davis and Pelsor 2001). We should understand this mechanism intuitively: it is the model of conventional agriculture. The farmer does not sow seeds into virgin prairie. Crops are planted into bare earth, then irrigated and fertilized. This is also the model of conventional site development. The developer clears the site to mineral soil, leaving only a few of the largest trees to be incorporated into the site design. "Landscaping" is installed, watered, and fertilized. Good for corn and wheat. Good for turfgrass. And good for invasive species.

Minimize bare ground during construction by enforcing tight limits of construction. Minimize trampling and vehicular use in areas not intended for high traffic and limit mowing of such areas to prescribed regimes (specifying the season, frequency, and mow height) intended to help achieve vegetation management goals. Enforce use policies (regarding off-trail use, dog leash rules, and the use of mountain bikes and other mechanized vehicles) on recreational trails to prevent vegetation damage and soil disturbance off the trail. Immediately close any unauthorized trails constructed by trail users.

LIMIT HABITAT FRAGMENTATION

Habitat edges are abrupt transitions between vegetation types—for example, where a forest abuts a neighborhood, or where a road bisects a prairie. (See Chapter 8 for more about habitat fragmentation.) Edges function as avenues for seeds of invasive species to enter a plant community. Thus, maximizing the area as well as the ratio of area to edge reduces the inflow of propagules into a habitat patch. In the absence of major human disturbance, many forests are able to resist invasion by exotic species (Brothers and Springarn 1992), but Levenson (1981) found that a forest area of at least 9.4 acres (3.8 hectares) was required to sustain forest-interior communities in the U.S. Midwest. The most powerful mechanism of invasion resistance in forests is low light availability (Brothers and Springarn 1992). Along forest edges, light levels are higher and invasions can be exacerbated.

Fragmentation similarly affects the invasibility of other ecotypes as well. For example, in the grasslands of the southern United States, roadsides facilitate the invasion of the grass King Ranch bluestem (*Bothriochloa ischaemum*) (Gabbard and Fowler 2007). Here, the grass is often seeded for erosion control. Maintenance of roadsides by frequent mowing further encourages invasion by suppressing competition from other tall grasses and distributing seed to new areas.

MAINTAIN HEALTHY DISTURBANCE

Plant and animal species and the ecosystems that they comprise evolved under a barrage of natural forces, such as fire, flooding, drought, herbivory, and predation. Humans have disrupted many of these forces or suppressed them completely, and ecological integrity has suffered as a result. A central strategy of ecological restoration and management is the reintroduction of natural disturbances that historically functioned as major ecological drivers or, failing that, the use of other techniques to mimic those disturbances. One benefit of this type of management is the control or reduction of some invasive species. For example, the use of prescribed fire during the growing season to mimic historic wildfire regimes has been demonstrated to be equally or more effective than chemical herbicides in controlling the invasive grass King Ranch bluestem (*Bothriochloa ischaemum*) (Simmons et al. 2007) (see Figure 7.10). As another example, the short-term lowering of water levels in pass-through reservoirs to mimic low-flow conditions during drought is used to control the aquatic invasive hydrilla (*Hydrilla verticillata*).

CITY OF AUSTIN

■ **FIGURE 7.10** A single-lane asphalt road divides a central Texas grassland that is dominated by the invasive grass King Ranch bluestem (*Bothriochloa ischaemum*). A prescribed burn was applied to the area on the right side of the road in September 2006 (growing season) and to the area on the left side of the road in January 2007 (dormant season). (The photo was taken in June 2007.) The growing-season fire significantly suppressed King Ranch bluestem and boosted native plant diversity, while the dormant-season fire improved productivity of native warm-season grasses and perennial forbs but failed to reduce King Ranch bluestem dominance.

Such strategies may be less practical on urban or suburban sites. On grassland or savanna sites, where grazing historically (i.e., presettlement) functioned as an ecological driver, prescribed mowing can be used to mimic the frequency and intensity of historic grazing regimes. For example, in North America, bison grazed very intensively but returned to areas infrequently. So in the grasslands of North America, prescribed mowing regimes should be infrequent but intense. The timing of mowing should coincide to remove flowering parts before the seeds of undesirable species mature or remove the aboveground tissue, particularly of woody plants, when belowground resources are limited (for example, just after spring green-up).

Furthermore, small sites should strive for high resistance and resiliency to natural disturbances (fire, flooding, drought, etc.) that are likely in the region or at the site. This will help to prevent ecological gaps in the wake of a disturbance that might be exploited by invasive species. Strategies include using vegetation that is resistant or resilient to anticipated disturbances. For example, in flood-prone areas, use native or site-appropriate vegetation that can thrive without supplemental irrigation during dry periods, as well as tolerate periods of inundation during flood events. In fire-prone areas, vegetation should integrate with the fire-protection strategies for the site and be highly resistant or resilient to wildfire. Such requirements may limit the plant palette for the site, but they will help protect the client's initial monetary investment, ensure the long-term stability of the site, and maintain inherent invasion resistance.

■ CASE STUDY

EVELYN PEASE TYNER INTERPRETIVE CENTER AT AIR STATION PRAIRIE

PROJECT TYPE: Public education

LOCATION: Glenview, Illinois

SIZE: 3,000 square-foot (279 square-meter) building on a 32-acre (13-hectare) site

COMPLETED: 2007

Highlighted Sustainable Practices:

- Integrated design process
- Zero stormwater runoff
- Removal and ongoing management of invasive species
- Protects and restores the natural hydrology of the site
- Engages the community in environmental education and stewardship activities

THE SITE: A 32-acre (13-hectare) open space parcel within the mixed-use redevelopment of the Glenview Naval Air Station. The master plan for the community included a high-quality remnant prairie, which is the location of an interpretive center (see Figure 7.11). Portions of the site were highly disturbed and had been filled with construction debris and other material. A new public roadway was constructed on the site's south and east borders, and a new commuter rail station is located a few blocks south.

■ **FIGURE 7.11**
Constructed on piers, the Evelyn Pease Tyner Interpretive Center avoids impeding groundwater flows and allows wetland vegetation to grow and recede during the rainy season. Aquatic landscapes were carefully created in the disturbed areas around the building to allow site visitors to experience the native wet prairie system up close.

Project Description

The Evelyn Pease Tyner Interpretive Center was built to provide educational opportunities and access to the Air Station prairie. The remnant prairie has been restored and is stewarded to maintain ecological health within an urban context. Natural processes, including periodic flooding and prescribed fire, have been restored across the site. Restoration efforts enhance the biodiversity and health of the prairie, which was threatened by the impacts of adjacent land uses and invasive species. As a result of the land management, local genotypes unique to this area are protected.

EVELYN PEASE TYNER INTERPRETIVE CENTER AT AIR STATION PRAIRIE *(CONTINUED)*

The educational facility includes a small, 3,000-square-foot (279-square-meter) structure that has open views to the surrounding landscape, as well as a series of boardwalks, trails, outdoor classrooms, and displays. The site provides an opportunity for visitors to experience the native plants, animals, and ecological processes of the region and develop a love for nature. Interpretive displays and literature, including self-guided tours and signage along the trail system, are integral to the educational mission.

Project goals included the protection and enhancement of the Air Station prairie and the provision of a leading-edge facility to serve as a model for sustainable building and infrastructure.

The development of the structure and pathways were used to restore stability and natural hydrology to the disturbed portions of the site. Vehicular access was kept along the road, with on-street parking and bus drop-off areas adjacent to the street, rather than an off-site parking lot, which would have been more disruptive and expensive. All of the new surfaces—green roof, porous pavements, and boardwalks—help to slow, cool, cleanse, and infiltrate rainwater, restoring natural hydrology to feed baseflow back to the prairie. As a result, there is no surface water runoff from the site.

The building was constructed on piers, to avoid impeding groundwater flows. Aquatic landscapes were created in the disturbed areas near the building so that it would appear as though it were placed in the middle of an existing wet prairie system.

The client, prairie stewards, and the landscape architect/ecologist led the site design process, which began with a thorough analysis of the site and surrounding area. A set of principles and initial concepts established the project objectives before an architect was retained. The design team was then expanded to include the architects, engineers, interpretive designers, and contractors. The collaborative design process included an ongoing series of design sessions and charrettes to evolve the design from concept through implementation. This close collaboration was essential to achieving the level of performance and design integration with the project.

© CONSERVATION DESIGN FORUM

■ **FIGURE 7.12**
The building orientation and architecture of the Evelyn Pease Tyner Interpretive Center reduce energy use by taking advantage of the southern winds and solar orientation. Interpretive features include demonstrations of the green roof system.

continues

EVELYN PEASE TYNER INTERPRETIVE CENTER AT AIR STATION PRAIRIE *(CONTINUED)*

The Interpretive Center is a model for sustainability and includes a green roof, water-conserving fixtures, geothermal heating and cooling, solar panels on a portion of the roof, and locally obtained, recycled products (see Figure 7.12). The project is Platinum LEED certified through the U.S. Green Building Council.

PROJECT TEAM

- CONSERVATION DESIGN FORUM, LANDSCAPE ARCHITECTURE, PLANNING, AND ECOLOGICAL RESTORATION
- DAVID YOCCA, FASLA, LEED AP, PROJECT PRINCIPAL
- PHOENIX ARCHITECTS, ARCHITECTURE, DESIGN, AND SUSTAINABILITY CONSULTANT
- WIGHT AND COMPANY, ARCHITECTURE AND CIVIL ENGINEERING
- LOIS VITT SALE, AIA, LEED AP, PRINCIPAL ARCHITECT
- JAY WOMACK, (FORMERLY WITH CDF AND WIGHT AND COMPANY), PROJECT MANAGER
- BLUESTONE + ASSOCIATES, INTERPRETIVE DESIGN
- PEPPER CONSTRUCTION, CONSTRUCTION MANAGER

Integrated Pest Management

Central to any strategic invasive species management strategy is an integrated pest management (IPM) plan. IPM is a method of pest control that relies on a combination of available, common-sense methods. An IPM plan identifies invasive species currently on-site as well as potential invaders, sets thresholds or trigger points for treatment actions, and prescribes methods for controlling established pests and preventing further spread and new introductions. The goal of IPM is to utilize the control methods that are both economical and present the least possible hazard to people, property, and the environment. The IPM plan is part of the site maintenance plan and is provided to the client upon completion of the project.

IDENTIFY PESTS

During site analysis, invasive species currently established on-site, as well as those not yet established but that have the potential to become established, should be identified. In determining which species to include in an IPM plan, project teams should utilize national and regional invasive species lists and consult professionals familiar with local invasive species issues.

SET ACTION THRESHOLDS

Not all invasive species have the same impacts or demand the same control treatments. Some species may only warrant action when they become very dense, whereas the single occurrence of another species may demand immediate removal. Some species may present realistic opportunities for complete removal, whereas for other, particularly pervasive species, significant control may not be feasible. Thresholds or trigger points indicate the level of infestation at which control actions should be taken. For residential or other small sites, complete eradication is likely to be possible and economical for many invasive species.

PREVENT THE ESTABLISHMENT OF NEW INVASIVE SPECIES OR THE SPREAD OF THOSE ALREADY ESTABLISHED

Use site-appropriate species. "Site-appropriate" is not necessarily inclusive of all native species—or exclusive of all nonnative species. Site-appropriate plants are those that are nonivasive and are able to thrive in the desired locations with minimal potable irrigation, no chemical fertilizers, and minimal fossil fuel–powered maintenance practices. Furthermore, nonnative species should not escape beyond the desired planting sites. If nonnatives begin to escape and naturalize in the surroundings, those species should be completely removed from the site.

Control the distribution of seed. Clean mowing and trimming equipment between sites to avoid transporting seeds of invasive species to new areas. Prescribed mowing may also be used to remove flower stalks before invasive plants produce seed. For annuals and biennials, this is sometimes an effective control strategy.

CONTROL

Effectively control invasive plants via best-management practices, passive management, and individual plant treatments such as cutting, pulling, and herbicide application.

REMOVAL METHODS

■ PASSIVE MANAGEMENT

Many invasive herbaceous species, particularly annuals and biennials, most often invade disturbed sites, such as agricultural fields, construction sites, and roadsides. Passive management uses dense perennial herbaceous vegetation to outcompete more diminutive, shorter-stature annuals and biennials, thereby reducing their dominance or entirely preventing them from establishing each year. This strategy only works when perennial vegetation is allowed to completely cover the soil and is left relatively undisturbed except by management treatments such as prescribed fire or prescribed mowing. The seeds of invasive species often remain in the soil for many years; therefore, any new soil disturbance or removal of perennial vegetation is likely to stimulate the return of the invasive annuals or biennials.

■ ACTIVE REMOVAL

When effective, manual or mechanical removal is preferred over chemical control. Hand-pulling can be effective when the full root system can be extracted from the ground. A weed wrench is useful for pulling shrubs and small trees with shallow root systems. Some woody species do not resprout when cut at the ground or top-killed. For these species, cutting, felling, mowing, or shredding is effective.

When passive management or mechanical removal is not available, herbicides can be an appropriate alternative. However, not all herbicides are equal in their effectiveness in controlling the target species, toxicity to nontarget plants and animals, or mobility in the soil, and not all herbicides are acceptable for use in or near water. Any herbicide, when improperly used, has the potential to cause harm to people, property, and the environment. Extreme care should be exercised when selecting and applying herbicides.

Select the least toxic, but effective, herbicide. Consider the following when selecting an herbicide:

- **SPECIMEN LABEL:** The specimen label is affixed to the original container in which the herbicide is sold and contains all immediately pertinent information, such as the concentration of the active ingredient(s); the actions to take in case of exposure; the required personal protective equipment (PPE); precautionary statements regarding the relative toxicity; directions for use, including the appropriate methods of application; and a list of pest species that the herbicide has been shown to control.
- **MATERIAL SAFETY DATA SHEET (MSDS):** The MSDS contains detailed information such as the potential health effects of exposure, impacts to wildlife and nontarget plants, and stability and reactivity. MSDSs can be obtained from the manufacturer.
- **HERBICIDE FACT SHEETS:** Fact sheets are publications—often peer-reviewed, produced, and distributed by third-party entities—and usually contain additional information not included in specimen labels or MSDSs, such as third-party research on human health effects and the environmental impacts of not only the active ingredient(s) but also the inactive, proprietary ingredients that often comprise the bulk of even "concentrated" herbicide formulations. Herbicide fact sheets are most easily found on the Internet.
- **REGULATORY STATUS:** In the United States, "general-use" herbicides may be purchased over-the-counter by the general public for personal use on private property. Others, because of an especially high risk to human health, the environment, or nontarget vegetation, may only be purchased and applied by a licensed pesticide applicator. These herbicides may be classified by the federal government as "Restricted Use" or by states as "Regulated" or "State-Limited-Use."
- **SELECTIVITY:** Nonselective herbicides kill a wide variety of vegetation. Selective herbicides are effective against a narrower range of plant types, such as broad-leafed plants or grasses.
- **EFFECTIVENESS FOR EACH SPECIES:** Herbicides vary in their ability to control various plants. The specimen label, third-party research, as well as experienced site managers may be able to indicate the most effective herbicide for a particular plant pest.
- **MIXING REQUIREMENTS:** The act of mixing concentrated herbicides with other substances prior to use increases the opportunity for accidental exposure or spills and introduces the possibility that the herbicide will be mixed incorrectly. Ready-to-use formulations typically do not require premixing beyond the addition of small volumes of dye or surfactants and reduce the potential for accidents. Ready-to-use formulations do, however, typically restrict the applicator to the use of more concentrated solutions.
- **ACTIVITY AND MOBILITY IN THE SOIL:** Some herbicides are bound by soil particles and have very low activity in soil, whereas others move freely through soil and groundwater and maintain their activity until they are biologically degraded.

MONITORING AND FOLLOW-UP

Though many invasive species can be effectively eradicated from small sites, periodic monitoring and follow-up treatment is always necessary to prevent reestablishment.

Invasive plants may reestablish any number of ways:

- Seeds may be stored in the soil.
- Seeds may be transported in from off-site.
- Individual plants may resprout after being cut or incompletely pulled or grubbed.
- Herbicide application may be ineffective for a portion of a treated population.
- Individual plants may simply be missed during a round of removal actions.

Monitoring and follow-up should occur on a schedule appropriate for the target species. Woody species and herbaceous perennials may require follow-up treatments every few years until populations fall below action thresholds, whereas annuals and biennials may need annual monitoring and follow-up treatment.

REFERENCES

Baker, H. G. 1974. The evolution of weeds. *Annual Review of Ecology and Systematics* 5: 1–24.

Brothers, T. S., and A. Springarn. 1992. Forest fragmentation and alien plant invasion of central Indiana old-growth forests. *Conservation Biology* 6: 91–100.

Burke, M. J. W., and J. P. Grime. 1996. An experimental study of plant community invasibility. *Ecology* 77: 776–90.

Clavero, M., and E. García-Berthou. 2005. Invasive species are a leading cause of animal extinctions. *Trends in Ecology and Evolution* 20: 110.

Colautti, R. I., S. A. Bailey, C. D. A. van Overdijk, K. Amundsen, and H. J. MacIsaac. 2006. Characterised and projected costs of nonindigenous species in Canada. *Biological Invasions* 8: 45–59.

Crooks, J. A., and M. E. Soulé. 2001. "Lag times in population explosions of invasive species: Causes and implications." In Sandlund, O., P. J. Schei, and Å Viken, eds., *Invasive Species and Biodiversity Management.* Dordrecht, Netherlands: Kluwer Academic Publishers.

Daehler, C. C. 2003. Performance comparisons of co-occurring native and alien invasive plants: Implications for conservation and restoration. *Annual Review of Ecology* 34: 183–211.

Davis, M. A., and M. Pelsor. 2001. Experimental support for a resource-based mechanistic model of invasibility. *Ecology Letters* 4: 421–28.

Dehnen-Schmutz, K., C. Perrings, and M. Williamson. 2003. Controlling *Rhododendron ponticum* in the British Isles: An economic analysis. *Journal of Environmental Management* 70: 323–32.

DiTomaso, J. M. 2000. Invasive weeds in rangelands: Species, impacts, and management. *Weed Science* 48: 255–65.

Dong, M., J.-Z. Lu, W.-J. Zhang, J.-K. Chen, and B. Li. 2006. Canada goldenrod (*Solidago canadensis*): An invasive alien weed rapidly spreading in China. *Acta Phytotaxonomica Sinica* 44: 72–85.

Executive Order no. 13112. "Invasive species." February 16, 1999. U.S. Federal Register 24 (25).

Gabbard, B. L., and N. L. Fowler. 2007. Wide ecological amplitude of a diversity-reducing invasive grass. *Biological Invasions* 9: 149–160.

Hobbs, R. J., and L. F. Huenneke. 1992. Disturbance, diversity, and invasion: Implications for conservation. *Conservation Biology* 6: 324–37.

Kennedy, T. A., S. Naeem, K. M. Howe, J. M. H. Knops, D. Tilman, and P. Reich. 2002. Biodiversity as a barrier to ecological invasion. *Nature* 417: 636–38.

Le Maitre, D. C., D. B. Versfeld, and R. A. Chapman. 2000 The impact of invading alien plants on surface water resources in South Africa: A preliminary assessment. *Water South Africa* 26: 397–408.

Le Maitre, D. C., B. W. van Wilgen, C. M. Gelderblom, C. Bailey, R. A. Chapman, and J. A. Nel. 2002. Invasive alien trees and water resources in South Africa: Case studies of the costs and benefits of management. *Forest Ecology and Management* 160: 143–159.

Levenson, J. B. 1981. "Woodlots as biogeographic islands in southeastern Wisconsin." In Burgess, R. L., and D. M. Sharpe, eds., *Forest island dynamics in man-dominated landscapes.* New York: Springer-Verlag. 13–39.

Lowe S., M. Browne, S. Boudjelas, and M. D. Porter. 2000. *100 of the world's worst invasive alien species: A selection from the Global Invasive Species Database.* Auckland, New Zealand: The Invasive Species Specialist Group (ISSG), a specialist group of the Species Survival Commission (SSC) of the World Conservation Union (IUCN): 12.

MacDougall, A. S., and R. Turkington. 2005. Are invasive species the drivers or passengers of change in degraded ecosystems? *Ecology* 86: 42–55.

Maron, J., and M. Marler. 2007. Native plant diversity resists invasion at both low and high resource levels. *Ecology* 88: 2651–61.

Marvier, M., P. Kareiva, and M. G. Neubert. 2004. Habitat destruction fragmentation, and disturbance promote invasion by habitat generalists in a multispecies metapopulation. *Risk Analysis* 24: 869–78.

Morin, N. 1995. "Vascular plants of the United States." In LaRoe, E.T., G. S. Farris, C. E. Puckett, P. D. Doran, and M. J. Mac, eds., *Our living resources: A report to the nation on the distribution, abundance, and heath of U.S. plants, animals, and ecosystems.* Washington DC: U.S. Department of the Interior, National Biological Service: 200–5.

Pimentel, D., R. Zuniga, and D. Morrison. 2005. Update on the environmental and economic costs associate with alien-invasive species in the United States. *Ecological Economics* 52: 273–88.

Pokorny, M. L., R. L. Sheley, C. A. Zabinski, R. E. Engel, T. J. Svejcar, and J. Borkowski. 2005. Plant functional group diversity as a mechanism for invasion resistance. *Restoration Ecology* 13: 448–59.

Rangelands NRM Western Australia. 2008. "Management of invasive species: Mesquite. Retrieved October 14, 2011, from http://www.rangelandswa.com.au/pages/147/management-of-invasive-species-mesquite.

Reinhardt, F., V. M. Herle, B. F. Bastiansen, and B. Streit. 2003. *Economic impact of the spread of alien species in Germany.* Berlin: Federal Environmental Agency.

Simberloff, D., and B. V. Holle. 1999. Positive interactions of nonindigenous species: Invasional meltdown? *Biological Invasions* 1: 21–32.

Simmons, M. T. 2005. Bullying the bullies: The selective control of an exotic, invasive annual (*Rapistrum rugosum*) by oversowing with a competitive native species (*Gaillardia pulchella*). *Restoration Ecology* 13: 609–15.

Simmons, M. T., S. Windhager, P. Power, J. Lott, R. K. Lyons, and C. Schwope. 2007. Selective and non-selective control of invasive plants: The short-term effects of growing-season prescribed fire, herbicide, and mowing in two Texas prairies. *Restoration Ecology* 15: 662–69.

Steiner, A. 2010. Counting the cost of alien invasions. BBC, April 13. http://news.bbc.co.uk/2/hi/science/nature/8615398.stm.

Suding, K. N., K. L. Gross, and G. R. Houseman. 2004. Alternative states and positive feedbacks in restoration ecology. *Trends in Ecology and Evolution* 19: 46–53.

Thompson, K., J. G. Hodgson, P. Grime, and M. J. W. Burke. 2001. Plant traits and temporal scale: Evidence from a 5-year invasion experiment using native species. *Journal of Ecology* 89: 1054–60.

Tinsley, M. J., M. T. Simmons, and S. Windhager. 2007. The establishment success of native versus non-native herbaceous seed mixes on a revegetated roadside in central Texas. *Ecological Engineering* 26: 231–40.

Williamson, M., and A. Fitter. 1996. The varying success of invaders. *Ecology* 77: 1661–66.

Zedler, J. B., and S. Kercher. 2004. Causes and consequences of invasive plants in wetlands: Opportunities, opportunists, and outcomes. *Critical Reviews in Plant Sciences* 23: 431–528.

ERICA NIX

■ FIGURE 8.1
An intensive green roof, designed and installed by Casey Boyter Gardens and Green Roofs, tops a small guesthouse in Austin, Texas. In addition to reducing stormwater runoff and providing superior building insulation, green roofs planted with a diverse community of native species can serve as valuable habitat for wildlife and plants. Seen blooming in the photo is devil's shoestring (*Nolina lindheimeriana*), which is endemic to the state of Texas. Also visible are yucca (*Yucca sp.*) (tall cream-colored inflorescence), zexmenia (*Wedelia texana*) (small yellow flowers), muhly (*Muhlenbergia sp.*), and other native grasses.

CHAPTER 8

Sustainable Solutions: Loss of Biodiversity

BY W. MATT MCCAW

PEOPLE GENERALLY BUY the idea that biodiversity is something to be saved and restored (Elder and Russonello 1997; Snaddon et al. 2008). The modern environmental movement has gotten this far. Unfortunately, the full meaning of *biodiversity* isn't always understood. We, the general public, think biodiversity happens in forests—especially rainforests. We think of fragile, imperiled, misty rainforests in foreign lands full of unseen, howling, wild things. We are told that fires and bulldozers and logging companies are bad and that they are killing biodiversity. And this is bad.

While the association between biodiversity and rainforests is appropriate (rainforests are, in fact, some of the most diverse natural systems on the planet), it is also overly simplistic, as is the idea that all fires and bulldozers and logging companies are bad. Biodiversity, in and of itself, is actually a fairly mundane concept, but the causes of its loss are highly complex and nuanced. Biodiversity is a characteristic of communities—even of the biological communities in and around our cities and towns. The biodiversity just outside our back doors provides the same benefits, to varying degrees, as that of the Amazonian jungle.

Biodiversity refers to the richness (number) and distribution (evenness) of species living in a given area. Highly diverse communities tend to contain a large number of species, many of which are common or occur frequently, whereas less diverse communities may have fewer species and/or may be dominated by just a few. Imagine a college campus with a diverse student body. It might be composed of students from twenty different ethnic backgrounds. Furthermore, none of these groups dominates campus culture, yet none is so uncommon that its influence is not felt. Now imagine a campus whose student body is less diverse. It might be composed of students from only five or six ethnic groups, or it might contain students belonging to the same twenty ethnic groups, as in our high-diversity campus, but one or two dominate the campus culture, so that the influence from the other eighteen or nineteen is scarcely felt.

The same notions of diversity hold true for biological systems. So when we talk about the loss of biodiversity, we're not just talking about extirpation or extinction as a result of human activity; we're also talking about the increasing dominance of a few generalist species as uncommon species become even more rare.

Loss of Biodiversity: The Cause

The design, construction, and maintenance of the spaces in which we live, work, and play can profoundly affect biodiversity. Common drivers of biodiversity loss associated with site development include the spread of invasive species, pollution, climate change, and the loss of habitat for plants and animals. All but the latter is discussed in other chapters; the bulk of this chapter will deal with strategies to effectively mitigate negative impacts to habitat and to restore damaged or degraded natural systems on-site.

Invasive Species

An invasive species is one that is nonnative to a particular ecosystem and whose introduction into that system causes or is likely to cause economic or environmental harm or harm to human health (Executive Order No. 13,112, 1999). Invasive species reduce biodiversity by displacing native species and/or altering the biophysical components of a landscape. In the worst cases, invasive species may displace most native species and completely dominate large areas. For a more detailed discussion of invasive species, see Chapter 7.

Pollution

The impact of pollution on biodiversity is most often a concern with regard to aquatic systems. Polluted runoff from landscapes, roads, and buildings degrades aquatic habitats and reduces the biodiversity of water bodies around the world. There are now hundreds of "dead zones" located at the mouths of major rivers, the most significant of which are located in the United States, Japan, and in European nations. In 1999, the dead zone at the mouth of the Mississippi River covered 7,700 square miles (19,940 km^2) (Joyce 2000). Dead zones arise when nutrients from fertilizer runoff, wastewater effluent, and fossil fuel emissions cause massive algal blooms. When the algae and other microorganisms die and fall to the ocean floor, their decay consumes dissolved oxygen in the water, the levels of which, over time, drop below what is required to sustain marine life. As a result, fisheries decline, and livelihoods suffer.

Climate Change

The earth's climate has been in flux since the crust began to cool and something novel first crawled up out of the goo. Until recently the climate has changed relatively slowly, and many species have been able to move around or adapt. As forests grew and withered, as the grassland biome emerged, as glaciers advanced and retreated, plants were able to cast seeds to more suitable ground. Even the most sedentary animal species had thousands of years to either translocate or evolve adaptations to cope with changing environmental conditions. Many hitched rides on nascent continents freshly cleaved from Pangaea.

There have indeed been at least seven major extinction events in our planet's history, but each of these played out over thousands of years (Benton 1995). Now our climate is changing more rapidly than ever before, and many species—perhaps 24 percent of species worldwide (Thomas et al. 2004)—may be unable to move or adapt fast enough to avoid extinction. Human-induced climate change may yet drive the most significant extinction event in the last sixty-five million years.

Habitat Loss

Habitat is the environment in which an animal or plant normally lives or grows. Habitat is often spoken of in very general terms (the desire to preserve or restore *habitat*, the value of *habitat*, etc.), without consideration of the species or groups of species that are being targeted. This generalization invites the question, "Habitat for what?" Habitat for a particular species (e.g., cut-throat trout)? Habitat for a certain suite of species (e.g., butterflies)? These are important questions to address, because virtually every landscape on this planet, no matter how degraded, is habitat for something. There is a giant, festering, toxic lake of acid near Butte, Montana, an abandoned open-pit copper mine, that is now habitat for a community of odd microorganisms. Desirable? No. But habitat nonetheless. Thus habitat loss does not literally mean the total loss of habitat for everything that previously lived in an area; it means the alteration of environmental conditions such that populations of certain species of interest are no longer sustained.

Habitat degradation refers to a reduction of habitat quality. Habitat quality is reduced when the breeding success of a particular species is compromised, when the number of individuals of a species that the habitat can support is reduced, or when the diversity of organisms supported by the habitat is reduced. Habitat may be degraded via the alteration of community components (species composition and diversity), structural components (physical habitat structure), or environmental components (climatic, chemical, and other abiotic variables such as pH or dissolved oxygen in aquatic habitats). Habitat may also be degraded via a reduction in habitat quantity (the aggregate area of habitat in an area), patch size (the size of a contiguous body of habitat), or connectivity.

On the other end of the loss-of-habitat spectrum is what is often referred to as habitat destruction: the dramatic and wholesale, often irreversible conversion of a natural system from a high-quality state to one that is much lower in quality and usually of a different ecotype entirely (e.g., the conversion of forest to lawn, prairie to parking lot, river to reservoir) (see Figure 8.2). Commonly, development practices result in the direct destruction of on-site habitats and the indirect degradation of off-site habitats via the propagation and spread of invasive species, the transport of pollution off-site, habitat fragmentation, and other mechanisms.

MATT MCCAW

■ **FIGURE 8.2**
A common window seat vista: grassland, woodland, and prime agricultural land recently converted to single-family residential housing. This site is located in southern Wisconsin.

Loss of Biodiversity: How It Affects Our Lives

Science has begun to elucidate some of the connections between biodiversity and human well-being. Presented here are but a few such connections, grouped into two categories.

The loss of biodiversity affects our lives in two ways:

1. By reducing the quality of ecosystem services
2. By removing individual indicators of the quality of ecosystem services

Reducing the Quality of Ecosystem Services

The quality of many services provided by natural systems is directly correlated to the level of biodiversity in those systems. For example, higher diversity grasslands and forests mitigate global climate change more effectively by sequestering atmospheric carbon at higher rates (Fornara and Tilman 2008; Chen 2006). In ecological restorations, high-diversity plantings improve species recruitment and help ensure the success of the project (Zedler et al. 2001). The psychological benefits of contact with nature have also been positively correlated with species diversity (Fuller et al. 2007). In addition, high diversity improves the resistance and resiliency of natural systems to large disturbances such as disease (Maron and Marler 2007), stand-replacing wildfire (White et al. 2009), and drought (Tilman and Downing 1994), which, in turn, protects vital services such as the mitigation of global climate change, provision of wildlife habitat, stabilization of stream banks, and provision of clean drinking water.

However, higher diversity is not always better. It is not necessary or desirable to augment diversity in an otherwise healthy system. The ultimate goal regarding the protection and restoration of biodiversity is the protection and restoration of the concomitant ecosystem functions. Recently disturbed sites are often characterized by high plant diversity. Oftentimes, these communities are dominated by undesirable natives or exotic species (Angold et al. 2006) and the quality of ecosystem services provided by these sites is often low. Diversity typically decreases slightly as systems mature and early-successional species are replaced by conservative perennials. Some systems—tidal marshes, for example—have naturally low diversity. They may sometimes be dominated by only half a dozen plant species (Zedler et al. 2001). But diversity is evaluated relative to what is considered normal in a healthy, intact system, so low plant diversity in tidal marshes is expected.

Removing Indicators of the Quality of Natural Support Systems

■ FIGURE 8.3 *Rhadine exilis*, an endangered ground beetle endemic to central Texas.

DR. JEAN K. KREJCA, ZARA ENVIRONMENTAL LLC

Every species, whether charismatic or obscure, is an indicator of the health of our natural support systems—a sentinel of environmental change. An example is a ground beetle called *Rhadine exilis* (see Figure 8.3), which has no common name. The U.S. Fish and Wildlife Service listed it and eight other so-called karst invertebrates as endangered in December of 2000 (U.S. FWS 2008). It is troglobitic, meaning that it lives its entire life underground. It is only known from forty-five caves in the eastern Edwards Plateau of central Texas. It lives in total darkness, primarily in cracks and small caverns that are too small to be accessed by humans. It is tiny—about 0.3 inches (7.4 mm) long. Scientists don't know exactly what it eats, but it likely depends on organic materials from the surface, such as leaf litter and animal droppings, washed underground. Only a few people have ever seen it, and these are mostly biologists who went looking for it. Even when biologists do find something they think might be *Rhadine exilis*, it is not readily identifiable; a specimen must be collected and sent off to a specialist who will examine it under a dissecting scope for confirmation. Obviously, *Rhadine exilis* is not driving ecosystem function on a large scale.

Now here is why we should care about *Rhadine exilis*: it lives within the limestone cracks and caves above the Edwards Aquifer, one of the most productive aquifers in the world (see Figure 8.4). The city of San Antonio, the seventh largest city in the United States, derives its entire municipal water supply from the Edwards. *Rhadine exilis* is sensitive to changes in the belowground environment. If it or any of the other rare karst invertebrates goes extinct, it will be because the belowground environment has deteriorated, has been plugged with sediments, paved over and cut off from the terrestrial world, or contaminated with pesticides or petrochemicals. If *Rhadine exilis* can no longer survive in the craggy voids of Edwards limestone, what does it foretell about the futures of the roughly two million people (Smith et al. 2005) who depend on the aquifer of the same name? In protecting karst invertebrates like *Rhadine exilis*, we protect our own quality of life.

DR. JEAN K. KREJCA, ZARA ENVIRONMENTAL LLC

■ **FIGURE 8.4**
Groundwater percolates deep below the surface to form the Edwards Aquifer in central Texas, dissolving calcium out of the Edwards limestone and redepositing it to form stalactites. Roughly two million people depend on the Edwards for water. Numerous insect species that live within the caves and smaller void spaces above the aquifer are listed by the U.S. Fish and Wildlife Service as endangered due to habitat destruction caused by filling, quarrying activities, and habitat degradation as a result of factors including altered surface water drainage patterns, changes in nutrient flow, invasive species, and pollution.

Such is the case throughout the world. If the ecosystems surrounding, and indeed within, our built communities support a vast array of plants and animals; if even the most sensitive aquatic species thrive in our natural water supplies; if we restore diverse rangelands that rebuild fertile soil and support higher stocking rates; if forests are more resistant to disease because of their high floristic diversity; if the spring flowers buzz with bees of all kinds, do we not ensure our own survival? If we can protect species throughout the world from an untimely demise, do we not protect ourselves from the same?

Protect and Restore Biodiversity

As complex as biological systems are, several key strategies will help project teams and site managers protect and restore biodiversity and maintain it over the long term:

- Appropriate site selection
- On-site habitat conservation
- Reduction of habitat fragmentation
- Restoration of ecosystem function
- Creation of small-patch habitats
- Holistic resource management
- Habitat mitigation

Appropriate Site Selection

Site location is one of the most important factors that determine a site's sustainability. It is important to select a site that not only meets the needs of the proposed project but also helps address local economic, social, and environmental problems. Projects should seek not merely to mitigate environmental damage but to improve overall environmental quality. New construction should be sited on previously developed land where ecosystem services have already been degraded. Greyfields and brownfields present the greatest opportunities for this kind of positive change. The redevelopment of greyfield or brownfield sites not only protects greenfields and reduces sprawl but also presents opportunities to restore natural systems and regenerate ecosystem services.

The development of greenfield sites should be a last resort, after opportunities to reuse previously built or degraded properties have been exhausted. However, not all greenfields are equal in their ecological value. Threatened ecological communities, habitat for threatened or endangered species, prime agricultural land, wetlands, floodplains, and groundwater recharge zones provide particularly valuable suites of ecosystem services. Negative impacts in these areas should be avoided even when legal mechanisms such as compensatory mitigation would allow a development to proceed.

On-Site Habitat Conservation

Healthy natural systems conserved from the outset of a project do not require resource-intensive restoration at the tail end. As simple as this concept is, it should frame the design team's approach to the site. Conserving healthy systems from the outset can help distinguish a project from other similar projects, translate into lower overall construction and management costs, and help maintain proper ecosystem functioning.

Strategies for conserving on-site habitat include:

- Incorporating existing vegetation and topography into the design solution.
- Locating new development on portions of the site that have already been degraded.
- Developing a site protection plan that minimizes clearing, grading, and other site disturbances.
  - Set adequate but tight limits of construction (LOCs). Delineate LOCs with fencing or other barriers and enforce these limits throughout construction.
  - Dictate monetary consequences for damage to areas or components of the site beyond the construction envelope. Make monetary consequences specific and calculable. For example, dollars per square inch of tree scar, dollars per square foot of soil disturbed, dollars per square foot of herbaceous vegetation damaged.
- Providing a specific location for equipment storage, material stockpiles, travel routes, and parking areas for construction equipment.
- Developing a plan for the safe use, storage, and transport of chemicals, fuels, and other hazardous materials.
  - Minimize the storage of these materials on-site. When on-site storage is necessary, identify appropriate holding areas and methods. Research local transportation and use regulations.
- Selecting construction equipment and practices to reduce damage to the site.
  - Avoid the use of oversize maintenance equipment that causes soil compaction and vegetation loss. Use the smallest and lightest tools that can effectively and economically accomplish the job.
  - Clean construction equipment between sites to prevent the spread of invasive species.

Reduction of Habitat Fragmentation

The geography of a site—its size, shape, and physical relationship to everything else around it—is critically important to habitat quality. The size and shape of a habitat patch determine the ratio of its area to its perimeter, or edge. The environmental conditions near the edge of a habitat patch are affected by the neighboring landscape. This is called edge effect. For example, for some distance inward, a forest edge is generally warmer, drier, and windier than the forest interior. The forest edge is, in essence, less like a forest than the interior. The edge of a habitat patch may also be affected by the influx of species from neighboring habitats. Thus, edge effect reduces habitat quality near the perimeter of a habitat patch and reduces the amount of core habitat.

The shape and size of a habitat patch play a role in determining its core area. By maximizing the ratio of habitat area to edge, the amount of core area is increased (see Figure 8.5).

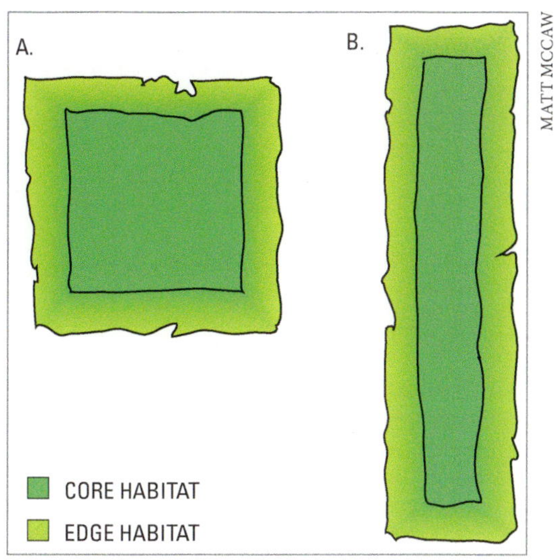

FIGURE 8.5
A square-shaped 15-acre (6-hectare) habitat patch has a perimeter of 3,233 feet (985 meters). Assuming a 100-foot (30-meter) edge effect, the area of core habitat is 8.5 acres (3.4 hectares) (a). An elongated 15-acre (6-hectare) habitat patch, in this case 404 feet (123 meters) by 1,633 feet (498 meters), has a perimeter of 4,074 feet (1242 meters). If the edge effect is 100 feet (30 meters), the area of core habitat will be reduced to 6.6 acres (2.7 hectares).

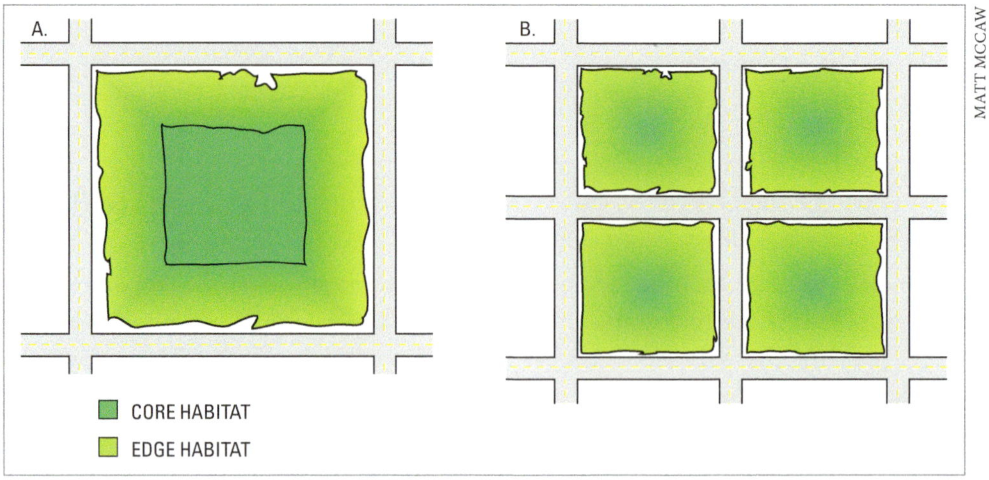

■ **FIGURE 8.6** Fragmentation increases edge and reduces core habitat area. One square-shaped 4-acre (1.6-hectare) habitat patch has a perimeter of 1,670 feet (509 meters) and, assuming a 100-foot (30-meter) edge effect, an area of core habitat of 1.1 acres (0.45 hectare) (a). By contrast, four square-shaped 1-acre (0.4-hectare) habitat patches have an aggregate perimeter of 3,339 feet (1,018 meters) and, assuming a 100-foot (30-meter) edge effect, essentially no core habitat (b).

The fragmentation of habitat is important in that it affects the flows of resources and the interactions between contiguous landscapes. The edge represents an additional point of entry for invasive nonnative or problematic native species. So fragmentation, by decreasing the ratio of habitat area to edge, decreases the core habitat area and increases the likelihood of invasion by undesirable species. This is important for, say, a wood thrush, an interior forest bird whose nests are predated by the brown-headed cowbird, a species that thrives in nearby agricultural lands and accesses other habitat types via edges. Thus, fragmentation harms wood thrushes by both decreasing the amount of suitable habitat and increasing the likelihood of nest predation by brown-headed cowbirds (see Figure 8.6).

Because of edge effect, the fragmentation of habitat is doubly harsh (see Figure 8.7). Fragmentation not only reduces total habitat area but also decreases the ratio of habitat area to edge. Development

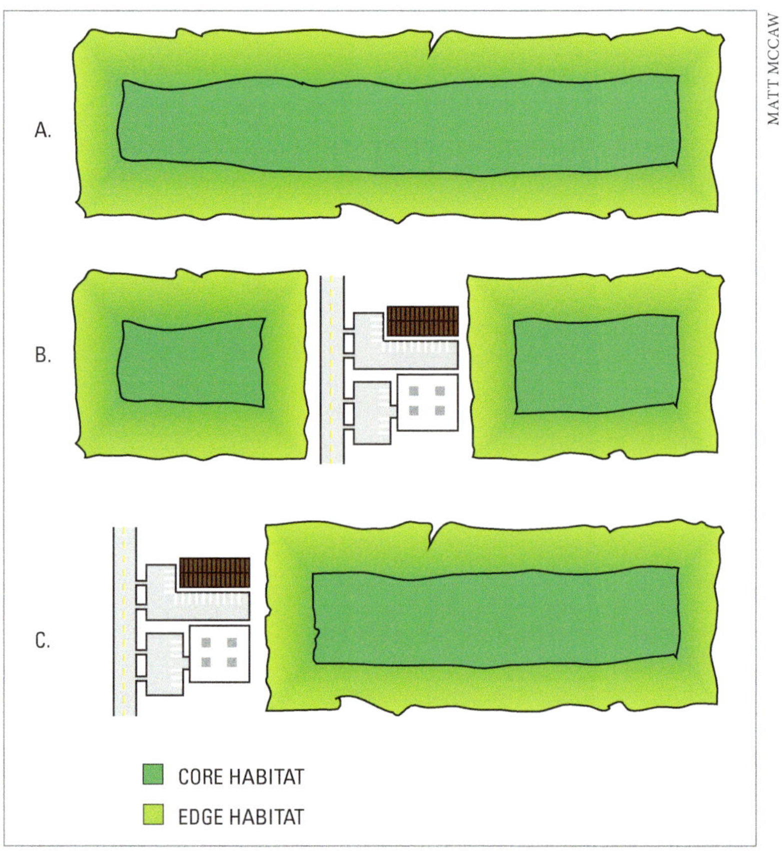

■ **FIGURE 8.7** The placement of a development that maintains a single habitat patch is preferable to one that splits a habitat patch in two. In this example, the original habitat patch is 11 acres (4.5 hectares). A 100-foot (30-meter) edge effect reduces the core habitat area to 4.6 acres (1.9 hectares) (a). A 3.7-acre (1.5-hectare) development that splits the habitat in two (b) reduces the core habitat to 1.8 acres (0.7 hectares). The same development sited on the edge of the patch, maintaining a contiguous body of habitat (c), preserves 2.8 acres (1.1 hectares) of core habitat.

actions that maintain habitat in large, uniform blocks may still reduce total habitat area, but they decrease the ratio of habitat area to edge to a lesser degree. Conversely, the restoration of habitat, when strategically sited, can be doubly beneficial. The strategic restoration of habitat can not only increase the total area of available habitat but also increase the ratio of habitat area to edge (see Figure 8.8).

Restoration of Ecosystem Function

Restoration is the process of assisting the recovery of an ecosystem that has been degraded, damaged, or destroyed (SER International 2004). The primary goal of a restoration project is rarely the re-creation of the pre-settlement landscape, but the recovery of ecological function and the provision of ecosystem services to improve the well-being of site users and the community at large. Small sites typically exist within the broader context of the human-dominated environment; restoration efforts on these sites can play an important role in improving ecosystem functioning on a much larger scale.

MATT MCCAW

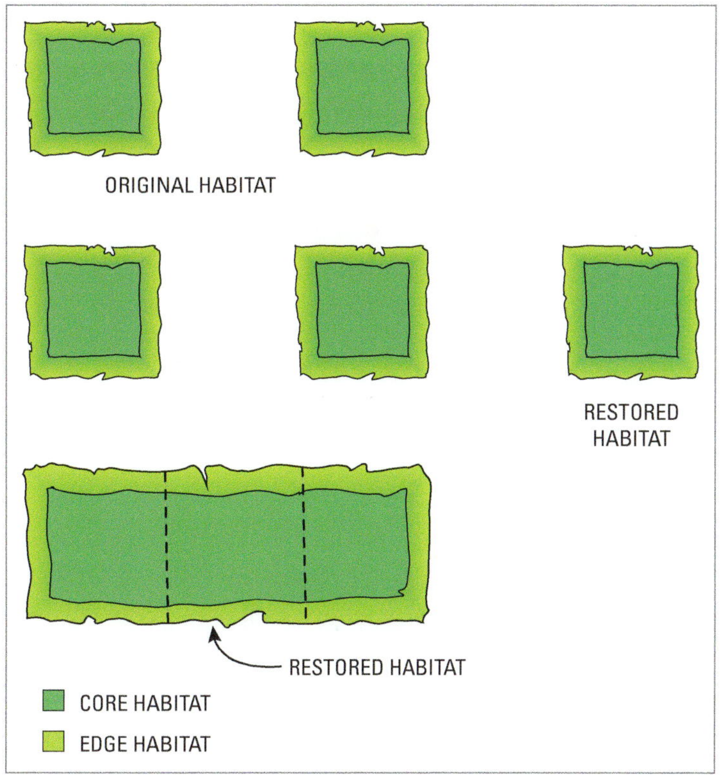

FIGURE 8.8
In this example, the original habitat (a) consists of two blocks, 5 acres (2 hectares) each. A 100-foot (30-meter) edge effect reduces the aggregate core habitat to 3.3 acres (1.3 hectares). An isolated 5-acre (2-hectare) block of restored habitat (b) results in 4.9 acres (2 hectares) of aggregate core habitat area. A contiguous 5-acre (2-hectare) block of restored habitat (c) connecting the two previously discontinuous blocks results in a single 7.3-acre (3-hectare) body of core habitat.

RESTORATION PLAN

Successful large restoration projects are guided by a well-researched and detailed plan. Even small restorations should be guided by a central document that communicates the overall intent of the restoration and the strategies for achieving success and addressing potential negative outcomes. A restoration plan should include:

- Statement of purpose
- Site description
- Goals and objectives
- Use policy
- Ecosystem models
- Implementation plan
- Management plan
- Monitoring plan

STATEMENT OF PURPOSE

The statement of purpose should outline the interests of the client and any other relevant background information that brought the client to that point.

If possible, the statement of purpose should identify target functions, that is, the highest priority ecosystem functions to be restored and maintained long-term. The most successful restorations are driven by one or two primary target functions—water quality, for example, or habitat for a particular species or a suite of species. Secondary target functions—perhaps carbon sequestration or outdoor recreation—are often included in the restoration plan, but restoration and management of the primary functions retains highest priority.

SITE DESCRIPTION

A site description should detail all relevant current conditions of the site, such as the current plant and animal species composition, including identification of desirable and undesirable species, soil types, hydrological conditions, and climatic conditions. Current and past land use, including anthropological or historical considerations, should be discussed. Current or future ecological drivers or outside influences, including the surrounding context and future development, should be considered.

PROJECT GOALS AND OBJECTIVES

Goals are general statements that help refine the direction of the project (for example, restoring native riparian vegetation). Objectives are measurable elements that help evaluate progress toward a specific goal. For example, an objective under the previous goal regarding riparian vegetation might be to eliminate all invasive woody species. The measurability of objectives is critical to success of the project. Objectives will integrate into the monitoring plan and allow monitoring data to inform the management of the site.

USE POLICY

Who (humans) will be the primary user groups? How and when will the site be used?

ECOSYSTEM MODELS

Ecosystem models are critical for formulating goals and objectives for restoration projects and for designing management and monitoring plans (SER International 2004). Ecosystem models detail important features of ecosystems and describe key interactions among those features. A good model will describe the physical, climatic, structural, functional, and biological components of an ecosystem and how those components may change over time in response to various ecological drivers. Ecosystem models should be backed by science, especially with respect to the ways in which biological components may change over time in response to potential management actions.

IMPLEMENTATION PLAN

The implementation plan outlines the logistics of carrying out the restoration. It should detail any site prep, soil amendments, or structural or engineered solutions; include or reference an integrated pest management plan for the control of undesirable species; specify methods for the introduction of desirable species; present an implementation schedule; and estimate costs (including inflation costs) for each element of the project.

MANAGEMENT PLAN

In some sense, restoration is a never-ending process. Ecosystems evolve and change over very long timescales, so it is unreasonable to assume that a fully functional natural system can be re-created in a few months or even years. Ecological management is the extension of restoration practices over decadal timescales. A management plan provides guidance after the original restoration team has moved on.

The management plan should facilitate adaptive management. It should detail the available treatment options for both short-term and long-term management, discuss alternatives to potentially ineffective management strategies, and propose contingencies for unintended negative consequences. It should also be revisited and revised on a schedule that parallels the data collection and analysis schedule put forth in the monitoring plan. This is the essence of adaptive management, completing the cycle from planning to monitoring to reevaluation of management actions.

The steps of adaptive management follow (see Figure 8.9):

1. Clarify the project mission, goals, and objectives.
2. Develop the monitoring and management plans based on the best available research and other information.
3. Begin implementing the monitoring plan to gather baseline data.
4. Begin implementing the management plan to initiate environmental change. Continue implementing the monitoring plan to gather information about the change occurring in the system.
5. Analyze monitoring data and evaluate the effectiveness of management actions relative to the project goals and objectives.
6. Adapt the management plan to improve effectiveness. Continue to update and refine the management plan with new research and other information.
7. Implement the new iteration of the management plan. Continue implementing the monitoring plan. Carry on the cycle of implement, monitor, evaluate, adapt.

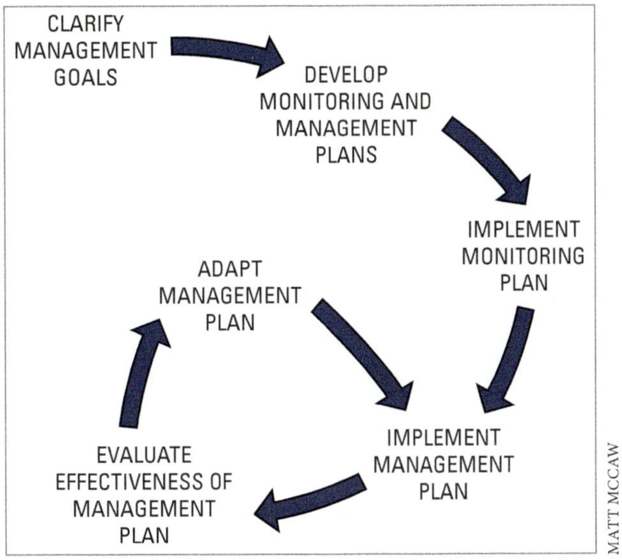

■ **FIGURE 8.9** The steps of adaptive management.

■ MONITORING PLAN

Monitoring is critical for the success of any initiative. In environmental management, good monitoring allows the ecosystem to communicate back to the land manager. The monitoring data should allow for evaluation of the progress toward the project's goals and objectives. The monitoring plan details the information to be collected, how and how often it will be collected, and how the data will be analyzed.

Creation of Small-Patch Habitats

Within the urban environment, where large, contiguous tracts of open space are rare, relatively high-diversity assemblages and even certain endangered species can be protected, restored, and sustained in small patches or in networks of small, unconventional habitat fragments (Angold et al. 2006), such as vegetated roofs, green walls, bioretention cells, rain gardens, woodlots, and stream corridors. For example, Kadas (2006) found that more than 10 percent of invertebrate species collected on vegetated roofs in London were designated nationally rare or scarce. In Berlin, Köhler (2006) observed 110 plant species over a twenty-year period, on ten vegetated roofs covering a composite area of only 6,997 square feet (650 square meters). Vegetated roofs are being specially designed in London as habitat for ground-nesting birds like the endangered black redstart and in San Francisco for the endangered butterfly, the bay checkerspot.

Small-patch habitats can provide important refuge sites for certain wildlife assemblages, particularly those with limited range requirements. For flying animals and aerially dispersed plants, linear corridors connecting larger habitat patches are not always vitally important for dispersal. For these groups, diversity and abundance are often affected more strongly by the quality and relative proximity of habitat patches than by habitat adjacency or habitat core–to-edge ratios; a patchwork of small habitat blocks permeating the landscape, serving as "stepping stones," is more critical for dispersal than the presence of linear habitat corridors connecting larger habitat patches (Angold et al. 2006) (see Figure 8.10). Larger animals (nonavian vertebrates), however, do frequently depend on habitat corridors connecting larger areas of contiguous habitat.

MATTIAS THOMA

■ FIGURE 8.10 The north portion of the Ackermannbogen Neighborhood Parkland is planted with ecologically appropriate vegetation, which is typical of the heath landscape around Munich, Germany. The dry meadow acts as a stepping-stone corridor for blue butterflies, partridges, and other endangered species.

DESIGN CONSIDERATIONS

The most significant challenge is often not protecting, constructing, or restoring a small habitat patch within an urban or residential setting, but maintaining habitat quality over the long term, especially in the face of criticisms that it looks weedy. The success of ecological design in urban settings often requires, in part, shifting the public's aesthetic sensibilities. Ecological quality tends to look messy. Our cultural expectations of how built landscapes should look often stymie our efforts to improve the way they function. According to Nassauer (1995), "What is good may not look good, and what looks good may not be good." Here, culture can be a barrier between good intentions to be green and holistic management of urban open spaces.

The care and intent in any urban landscape, be it a wildflower planting, a wetland, or a prairie restoration, must be communicated clearly to the public. Many people may perceive unmowed land as weedy, neglected, or even dangerous. Further, if that seemingly neglected land is public, these perceptions may reflect negatively upon the agency charged with its care. Successful "cues to care" have been demonstrated around the world and include edging, limited mowing along sidewalks and roadways (See Figure 8.11), invasive species management, wildflower plantings, birdhouses and bird feeders, and signage indicating that such areas are being managed for certain characteristics (Nassauer 1995) such as wildflowers, water quality, or carbon sequestration.

MATT MCCAW

■ **FIGURE 8.11** A narrow mowed strip along a crushed granite pedestrian trail in Austin, Texas communicates the intent of the adjacent restored tallgrass prairie vegetation.

Holistic Resource Management

Natural systems do not exist in steady states, but continually respond to and recover from disturbances caused by natural forces such as fire (see Figure 8.12), flooding, drought, herbivory, predation, and disease. Without such disturbances, natural systems may transition from one ecological state to another, (e.g., from a grassland to a woodland), may lose diversity and ecological integrity over time (Rogers et al. 2008), or be colonized by invasive species (Hobbs and Huenneke 2002). Natural areas must be actively managed in order to sustain ecological integrity and ecosystem services. Management strategies should reintroduce or mimic natural disturbances.

CITY OF AUSTIN

■ **FIGURE 8.12** Seep muhly (*Muhlenbergia reverchonii*) resprouts just weeks after a wildfire west of Austin, Texas. Land managers often use prescribed burns to re-create the effects of wildfire as a key strategy for restoring and maintaining desired habitat structure and diversity.

Certain management strategies, such as grazing or prescribed burning, will not be possible in all areas. In such cases, alternatives that offer similar outcomes should be sought. For example, depending on the project goals and the scale of the operation, prescribed mowing coupled with selective removal of woody plants and other undesirable species is sometimes a workable alternative to prescribed burning or grazing. In other cases, certain strategies may require outreach, education, facilitation, or other means in order to gain acceptance from the community. For example, many people respond negatively to the removal of trees but may support the removal of invasive trees if they understand the justifications and the desired outcomes.

All management strategies should be holistic, a term that has unfortunately become a bit clichéd and may carry different meanings in different contexts. Regarding the management of natural areas and the provision of ecosystem services, *holistic* refers to the management of all components of a natural system. Just as the members of an ecological community are connected, the services provided by the community are connected as well. An improvement in one service or function is met by improvement in many of the other services and functions. Thus, a holistic management strategy will improve or sustain the first-priority ecosystem service to the betterment of the majority of the other ecosystem services provided by the site. In contrast, a nonholistic management strategy will improve or sustain the first-priority ecosystem service to the detriment of the majority of the other services provided on-site.

Holistic management strives for optimization, rather than maximization, of the first-priority service. As an example, in a city park, a holistic approach might strive to maintain relatively high park use rates (the first-priority service) but would also manage second-priority ecosystem services, such as water quality, biological diversity, and mitigation of the urban heat island effect, and would manage park use or restrict access to certain areas of the park in order to prevent degradation of these services. In contrast, a nonholistic approach would maximize park use as other ecosystem services degrade.

Habitat Mitigation

Mitigation is a decision-making framework that seeks to eliminate the negative impacts of development actions by avoiding or minimizing negative impacts, restoring damaged environments, and/or compensating for negative impacts to habitat or other natural resources such as wetlands or threatened plant communities (OTA 1984).

In the traditional mitigation process:

1. Avoidable negative impacts are avoided.
2. Impacts that cannot be avoided are minimized.
3. When impacts cannot be avoided or minimized, the damaged environments are restored.
4. When negative impacts are unavoidable, the lost resources or functions are replaced or comparable substitutes are provided.

Within some regulatory environments, mitigation may refer specifically to the fourth step in the mitigation process, called compensatory mitigation, which involves the protection, restoration, or creation of off-site resources in exchange for anticipated on-site resource losses. A challenge with this mechanism is ensuring that the mitigation actions more than compensate for the losses.

It is the rare case when compensatory mitigation is in keeping with the spirit of sustainability. Frequently compliance is achieved by the purchase of credits in a habitat conservation bank (see Figure 8.13) or wetland bank. Compensatory mitigation often results in a net loss of either overall quantity or quality of habitat, or both.

■ **FIGURE 8.13** Downtown Austin, Texas, seen from one of the tracts of the Balcones Canyonlands Preserve, a habitat conservation bank established by the Balcones Canyonlands Conservation Plan (BCCP) to mitigate loss of habitat for eight endangered species. When negative impacts to habitat cannot be otherwise minimized or eliminated, landowners and developers may elect to mitigate development of endangered species habitat by participating in the BCCP on a fee basis, rather than mitigating directly through the U.S. Fish and Wildlife Service.

In the case of wetlands, the no-net-loss policy in the United States allows for the "conversion" (i.e., destruction) of wetlands by development actions if developers restore or create an equal or greater area of wetlands elsewhere, oftentimes by constructing new wetlands. The recommendations that encouraged this policy, the National Wetlands Policy Forum (1988), and the agencies responsible for implementing it (the Army Corps of Engineers and the Environmental Protection Agency) indicate that *functions* shall be preserved. Most often, however, because of our sheer inability to re-create natural systems from scratch, the full functionality of the converted wetland cannot be replicated (NRC 1992; Zedler 1996). Thus, preserving wetland function along with wetland area can be especially difficult. Restoring degraded wetlands as mitigation, as opposed to creating new, inferior wetlands, is one strategy for preserving net wetland function via compensatory mitigation.

In the case of endangered species habitat, the U.S. Endangered Species Act allows for the "take" (i.e., destruction or degradation) of habitat for threatened or endangered animals (endangered plants have no legal protection in the United States) if an equal or greater amount of habitat is preserved or created elsewhere. Perhaps the most common avenue for achieving compliance is the purchase of credits in a habitat mitigation bank. In the end, habitat is protected in one location (the mitigation bank) but destroyed or degraded in another location (the development site). Diligent management to improve the quality of the mitigation habitat is necessary to prevent a net loss of the total available habitat over time.

Appropriate mitigation ratios can also help improve the sustainability of mitigation projects. A mitigation ratio is the area of a created, restored, or protected resource that must be provided in exchange for a given area of degraded or destroyed resource. For example, a one-to-one mitigation ratio would require one unit of mitigated habitat in exchange for one unit of degraded or destroyed habitat. Depending on the quality of the replacement habitat, the quality of the degraded habitat, and the degree of degradation, higher mitigation ratios—three or more-to-one—may be necessary to meet the intent of preserving quantity and function.

From a sustainability standpoint, the preferable alternative, except in extraordinary circumstances, is the removal of compensatory mitigation as an acceptable option. Net-negative impacts that cannot be avoided, minimized, or rectified are a sign that a development action is inappropriate on the given site. In this case, the project should be altered to avoid net-negative impacts or alternative sites should be considered.

When a design team is not in a position to dramatically alter a project plan or select an alternative site, compensatory mitigation may be the last best alternative for addressing negative environmental impacts. These guidelines can help ensure the quality and integrity of a mitigation project:

- Work with established conservation initiatives such as land trusts, regional habitat conservation plans, or species recovery plans when mitigating habitat or wetland losses.
- Purchase credits in a habitat or wetland bank that has been endorsed by the regulatory agency requiring mitigation or a prominent conservation organization (e.g., the World Wildlife Fund, Conservation International, The Nature Conservancy) and is part of a broader conservation initiative. Work with regulatory agencies and conservation organizations to ensure the legitimacy, quality, and integrity of mitigation projects.
- Work with regulatory agencies and conservation organizations to ensure appropriate mitigation ratios.

■ CASE STUDY

TANNER SPRINGS PARK

LOCATION: Portland, Oregon

SIZE: 1.2 acres (0.5 hectare)

COMPLETION DATE: 2005

CLIENT: City of Portland

HIGHLIGHTED SUSTAINABLE PRACTICES:

- Brownfield redevelopment
- Extensive stakeholder process
- Increased vegetative biomass
- Reduced input to existing storm drain systems
- Stormwater management features beautifully integrated into the site to encourage interaction with nature

HERBERT DREISEITL / © ATELIER DREISEITL

■ **FIGURE 8.14**
More than three hundred citizens were involved in three public events at which art, brainstorming, and planning workshops informed and inspired the design process.

THE SITE: The brownfield site located in the Pearl District is a historic wetland that was filled to make way for warehouses and rail yards. Over the past thirty years, a progressive and dynamic mixed-use neighborhood has established itself, and today the Pearl District is home to families, businesses, and public spaces. Most recently, the site was used as a staging area for construction and had no existing vegetation.

Design Overview

Tanner Springs is an urban water park that provides green space in a previously industrial area and reconnects visitors with nature. The project seeks to capitalize on the sensory characteristics of a wetland while embracing the urbanity of the surrounding mixed-use neighborhood. More than three hundred citizens were involved in three public events at which art, brainstorming, and planning workshops informed and inspired the design process (see Figure 8.14).

TANNER SPRINGS PARK *(CONTINUED)*

The purpose of the park is not to restore a native wetland but to use natural processes similar to those found in a wetland to cleanse and manage on-site stormwater. Combined sewer overflows during wet weather occur in the Willamette River, which runs through Portland, an average of one hundred days per year. In response to this environmental and human health issue, Atelier Dreiseitl created a living water system that reduces inputs to existing storm drain systems.

All stormwater runoff from the 1.2-acre (0.5-hectare) site flows to the cleansing biotope and lower pond at the eastern end of the property (see Figure 8.15). The biotope, comprised primarily of coarse sand and plant media, functions as a wetland and supports native vegetation that begins the cleansing process. After moving through the soil and vegetation, water is treated with ultraviolet light via an underground utility vault, then pumped to the man-made springs at the top of the slope. The water then forms streams that are accessible to park visitors and slowly meanders through the site back to the biotope. Five-year storm events are managed on-site; additional stormflow is sent to the public storm drain.

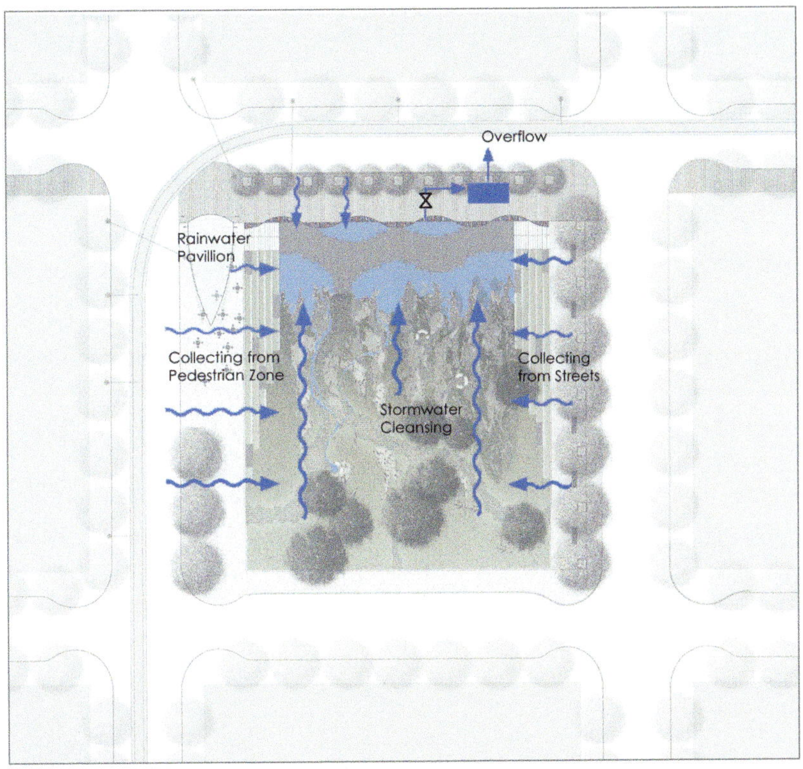

■ **FIGURE 8.15**
Stormwater management conceptual diagram.

Native vegetation covers the majority of the site and includes trees obtained by an Oregon tree salvage company (see Figure 8.16). Similar to a natural ecosystem, the vegetation in the biotope is intended to be self-selecting based on growing conditions. Symbolic of the old city fabric, historic railroad tracks form a wave-wall along the edge of the lower pond. The "Art Wall" acts as a visual backdrop and barrier to the noise and commotion of the surrounding city. It is 60 meters (197 feet) long and composed of 368 rails, with 99 pieces of fused glass inset with images of nature hand-painted by artist Herbert Dreiseitl. Bleacherlike lawn terraces provide a place for leisure and a connection to the water's edge.

continues

TANNER SPRINGS PARK *(CONTINUED)*

■ **FIGURE 8.16**
Native vegetation covers the majority of the site. Similar to a natural ecosystem, the vegetation in the biotope is intended to self-select in response to growing conditions.

During construction, temporary natural fences made of red twig dogwood cuttings protected the wetland plantings. The site is a dog-free park, as animals can damage the biotope. To address demand, the city established a grassy area for dogs one block north of the site, which has helped local residents accept the park's restrictions.

PROJECT TEAM

- **ATELIER DREISEITL, LANDSCAPE ARCHITECTS**
  www.dreiseitl.com
- **GREENWORKS, PC, LANDSCAPE ARCHITECTS**
  www.greenworkspc.wordpress.com

REFERENCES

Angold, P. G., J. P. Sadler, M. O. Hill, A. Pullin, S. Rushton, K. Austin, E. Small, et al. 2006. Biodiversity in urban habitat patches. *Science of the Total Environment* 360: 196–204.

Benton, M. J. 1995. Diversification and extinction in the history of life. *Science* 268 (5207): 52–8.

Chen, X. 2006. Tree diversity, carbon storage, and soil nutrient in an old-growth forest at Changbai Mountain, Northeast China. *Communications in Soil Science and Plant Analysis* 37: 363–75.

Elder, J., and J. Russonello.1997. "Reaching the public: The challenge of communicating biodiversity. In Raven, P. H., and T. Williams, eds. *Nature and human society: The quest for a sustainable world.* (Proceedings of the Forum on Biodiversity.) Washington, DC: National Academy Press.

Executive Order no. 13,112. "Invasive species." February 16, 1999. *U.S. Federal Register* 24 (25).

Fornara, D. A., and D. Tilman. 2008. Plant functional composition influences rates of carbon and nitrogen accumulation. *Journal of Ecology* 96: 314–22.

Fuller, R. A., K. N. Irvine, P. Devine-Wright, P. H. Warren, and K. J. Gaston. 2007. Psychological benefits of greenspace increase with biodiversity. *Biology Letters* 3: 390–94.

Hobbs, R. J., and L. F. Huenneke. 1992. Disturbance, diversity, and invasion: Implications for conservation. *Conservation Biology* 6: 324–37.

Joyce, S. 2000. The dead zones: Oxygen-starved coastal waters. *Environmental Health Perspectives* 108: A120–25.

Kadas, G. 2006. Rare invertebrates colonizing green roofs in London. *Urban Habitats* 4: 66–86.

Köhler, M. 2006. Long-term vegetation research on two extensive green roofs in Berlin. *Urban Habitats* 4: 3–26.

Maron, J., and M. Marler. 2007. Native plant diversity resists invasion at both low and high resource levels. *Ecology* 88: 2651–61.

Nassauer, J. I. 1995. Messy ecosystems, orderly frames. *Landscape Journal* 14: 161–70.

National Research Council [NRC]. 1992. *Restoration of aquatic ecosystems: Science, technology, and public policy.* Washington, DC: National Academy Press.

National Wetlands Policy Forum. 1988. *Protecting America's wetlands: An action agenda.* Washington, DC: Conservation Foundation.

Office of Technology Assessment [OTA]. 1984. *Wetlands: Their use and regulation.* Washington, DC.

Rogers, D. A., T. P. Rooney, D. Olson, and D. M. Waller. 2008. Shifts in southern Wisconsin forest canopy and understory richness, composition, and heterogeneity. *Ecology* 89: 2482–92.

Smith, B., B. Hunt, I. Jones, R. Lindgren, and G. Schindel. 2005. Ground-water availability modeling of the three major segments of the Edwards Aquifer, central Texas. Groundwater Summit, National Ground Water Association, April 2005, San Antonio, TX.

Snaddon, J. L., E. C. Turner, and W. A. Foster. 2008. Children's perceptions of rainforest biodiversty: Which animals have the lion's share of environmental awareness? *PLoS ONE* 3: e2579.

Society for Ecological Restoration [SER] International Science and Policy Working Group. 2004. *The SER International primer on ecological restoration.* Tucson: Society for Ecological Restoration International.

Thomas, C. D., A. Cameron, R. E. Green, M. Bakkenes, L. J. Beaumont, Y. C. Collingham, B. F. N. Erasmus, et al. 2004. Extinction risk from climate change. *Nature* 427: 145–48.

Tilman, D., and J. A. Downing. 1994. Biodiversity and stability in grasslands. *Nature* 367: 363–65.

U.S. Fish and Wildlife Service [FWS]. 2008. Bexar County karst invertebrates draft recovery plan. Albuquerque, NM: U.S. Fish and Wildlife Service.

White, J. D., J. Thomas, D. Murray, M. Sides, and J. Yao. 2009. The Balcones Canyonlands Preserve fire risk and management: Characterization of woodland fuels and simulated fire behavior in the wildland-urban interface. Report presented to the City of Austin Wildland Conservation Divison.

Zedler, J. B. 1996. Ecological issues in wetland mitigation: An introduction to the forum. *Ecological Applications* 6: 33–7.

Zedler, J. B., J. C. Callaway, and G. Sullivan. 2001. Declining biodiversity: Why species matter and how their functions might be restored in Californian tidal marshes. *Bioscience* 51: 1005–17.

Index